Cambridge Scientific M...

Since the 'scientific revolution' of the seventeenth century, a great number of distinguished scientists and mathematicians have been associated with the University of Cambridge. *Cambridge Scientific Minds* provides a portrait of some of the most eminent scientists associated with the University over the past 400 years, including accounts of the work of three of the greatest figures in the entire history of science, Isaac Newton, Charles Darwin, and James Clerk Maxwell.

The chronological balance reflects the increasing importance of science in the recent history of the University. The book comprises personal memoirs and historical essays, including contributions by leading Cambridge scientists. *Cambridge Scientific Minds* will be of interest not only to graduates of the University, science students, and historians of science, but to anyone wishing to gain an insight into some of the greatest scientific minds in history.

PETER HARMAN is Professor of the History of Science at Lancaster University. At Cambridge he lectured in the history of science and was a Research Fellow of Clare Hall. He has written several books published by Cambridge University Press.

SIMON MITTON studied Physics at the University of Oxford, following which he undertook a Ph.D. in radio astronomy at the Cavendish Laboratory, University of Cambridge. He is Senior Fellow of St Edmund's College, Cambridge. This is his tenth book.

Cambridge Scientific Minds

Edited by

Peter Harman and Simon Mitton

PUBLISHED BY THE PRESS SYNDICATE OF THE UNIVERSITY OF CAMBRIDGE
The Pitt Building, Trumpington Street, Cambridge, United Kingdom

CAMBRIDGE UNIVERSITY PRESS
The Edinburgh Building, Cambridge CB2 2RU, UK
40 West 20th Street, New York NY 10011-4211, USA
477 Williamstown Road, Port Melbourne, VIC 3207, Australia
Ruiz de Alarcón 13, 28014 Madrid, Spain
Dock House, The Waterfront, Cape Town 8001, South Africa

http://www.cambridge.org

© Cambridge University Press 2002

This book is in copyright. Subject to statutory exception
and to the provisions of relevant collective licensing agreements,
no reproduction of any part may take place without
the written permission of Cambridge University Press.

First published 2002
Reprinted 2002

Printed in the United Kingdom at the University Press, Cambridge

Typeface Trump Mediaeval 9.5/15 pt.   *System* QuarkXPress™   [SE]

A catalogue record for this book is available from the British Library

ISBN 0 521 78100 0 hardback
ISBN 0 521 78612 6 paperback

Contents

	Foreword by Sir Alec Broers	*page* vii
	Introduction PETER HARMAN	1
1	William Gilbert STEPHEN PUMFREY	6
2	William Harvey ANDREW CUNNINGHAM	21
3	Isaac Newton: Creator of the Cambridge scientific tradition RUPERT HALL	36
4	William Whewell: A Cambridge historian and philosopher of science RICHARD YEO	51
5	Adam Sedgwick: A confident mind in turmoil DAVID OLDROYD	64
6	Charles Babbage: Science and reform ANTHONY HYMAN	79
7	Charles Darwin PETER J. BOWLER	94
8	Stokes and Kelvin, Cambridge and Glasgow, light and heat DAVID B. WILSON	107
9	James Clerk Maxwell SIMON SCHAFFER	123
10	The duo from Trinity: A.N. Whitehead and Bertrand Russell on the foundations of mathematics, 1895–1925 IVOR GRATTAN-GUINNESS	141

11	Thomson, Rutherford and atomic physics at the Cavendish BRIAN PIPPARD	155
12	Hopkins and biochemistry HARMKE KAMMINGA	172
13	Charles Sherrington, E.D. Adrian, and Henry Dale: The Cambridge Physiological Laboratory and the physiology of the nervous system E.M. TANSEY	187
14	Hardy and Littlewood ROBIN J. WILSON	202
15	Arthur Stanley Eddington MALCOLM LONGAIR	220
16	Paul Dirac: A quantum genius HELGE KRAGH	240
17	Alan Turing ANDREW HODGES	253
18	Francis Crick and James Watson ROBERT OLBY	269
19	Mary Cartwright TOM KÖRNER	282
20	Joseph Needham GREGORY BLUE	299
21	Molecular biology in Cambridge M.F. PERUTZ	313
22	The discovery of pulsars – prelude and aftermath ANTONY HEWISH	325
23	Stephen W. Hawking SIMON MITTON	338

Foreword

For much of its almost 800-year history, Cambridge University's greatest minds devoted themselves to fields other than science as we know it. As this book shows, it was not until the time of William Gilbert in the late sixteenth century, already three centuries into the University's life, that science in our sense began to characterise its life and work.

Of course, as we see here, the foundations existed, especially in the mathematical studies, which later came to underlie the whole Cambridge curriculum and provided the soil in which science was to be planted and grow. Isaac Newton in particular, who shines in the firmament of Cambridge stars, continues to inspire our mathematical and physical studies, not least in the new Institute which bears his name.

In 1664, in Newton's prime, the new Royal Society of London took as its motto 'Nullius in verba' ('on the word of no man'), signifying that evidence, tangible data, and experience repeated at will were to be the marks of scientific endeavour. Merely to quote opinion or authority was in future to be valueless. And so, from then, we move into what is increasingly recognisable as the modern world of scientific inquiry, and pass to the era of Cambridge's greatest contributions to our knowledge of the world.

From then until our own day the progress of Cambridge and of the World's science studies is clear, as these chapters skilfully describe. They are themselves mainly the work of current scientists and historians of science, who bring their own insight to bear on their subjects. And, if we know little about the lives and characters of the earliest people to appear in this book, all the more welcome and valuable are the autobiographical notes of some of our contemporaries who carry on the great tradition.

If I have one plea, though, it is this. The contribution that Cambridge has made to the amelioration of suffering, the improvement of living conditions, the widening of knowledge and experience is incalculable. But we should not let this blind us to the rest of our University's activity. In this uncertain and often threatening world, there has never been a greater need for the humane, artistic and cultural studies which enhance our lives and also form a vital part of our curriculum.

Professor Sir Alec Broers FRS FREng
Vice-Chancellor
Cambridge, October 2001

Introduction
PETER HARMAN

The University of Cambridge can point to a long roster of distinguished scientists and mathematicians who have been associated with its history over the past 400 years, since the 'scientific revolution' of the seventeenth century. Their names include three of the greatest luminaries in the entire history of science: Isaac Newton, Charles Darwin, and James Clerk Maxwell. The link between the Cambridge present and the magnitude of past achievement is highlighted by the names given to institutions and buildings in the University: Darwin College, the Newton Institute, Harvey Court. This sense of continuity, that the work of the present age is linked to the traditions of the founding fathers, has prompted this collection of essays on Cambridge scientists, intended to interest a broad non-specialist readership within and beyond the University. But to the historian such associations, though beguiling, are problematic, suggesting sentiment or constructed 'heritage' rather than historical reality. Can figures of the early modern period – Isaac Newton, William Gilbert, William Harvey – be properly described as 'scientists'? The term, with its resonances of professional specialisation, was only introduced (by another Cambridge notable, William Whewell) in 1834. Does Cambridge 'science' have a 400-year history, or is it a product of the professionalisation of science, of teaching and research, since 1850 and especially since 1900? Is there indeed a specifically Cantabrigian scientific culture? The three major scientific figures associated with the University illustrate some of the problems in defining Cambridge science. Darwin was educated at Cambridge as was Maxwell, who returned later in his career to undertake the foundation of the Cavendish Laboratory, a development which was fundamental to the history of the University. But both Darwin and Maxwell carried out their main scientific work away from the University; only

Newton had strong association with Cambridge through his education, science, and career.

Though 'Cambridge science' can be seen to have shifted in meaning, as teaching and research in Cambridge changed, the term does have clear associations. Cambridge is widely seen as a scientific university, its culture shaped by a matrix of tradition and current practice. The essays in this book explore these scientific associations. In its provision of science, throughout its history, the University responded to pressures from outside the University, from government or from the changing aspirations of society at large. The place of science at Cambridge needs to be viewed in relation to the wider history of science since 1600, and to the changing patterns of education as the University adjusted to the demands of the modern world. While the essays in the present volume are not written to any set pattern, they illustrate these developments by locating Cambridge scientists within their historical milieux. The intention here is to illustrate the diversity of Cambridge associations; not to claim scientists (who may have done their major work elsewhere) for Cambridge, but to recognise the links to the past as meaningful, not mere sentiment. The tradition of past associations and achievements has helped shape the place of science within the culture of Cambridge. Delivering the University's annual Rede Lecture in 1878, James Clerk Maxwell could make passing reference to 'one of our own prophets – William Harvey, of Gonville and Caius College'. In the 1870s science was beginning to emerge as a significant element in the University's activities, and Maxwell sought to foster its integration into the humanistic culture of Cambridge. But the great flowering of science at Cambridge is a phenomenon of the twentieth-century.

The attainment by the sciences of appropriate methods of enquiry and social institutions is a feature of their recent history. Since the emergence of natural science in the Renaissance as a seminal feature of European intellectual life three major periods of scientific change stand out. First, the intellectual revolution of the seventeenth century in which the cognitive basis of modern natural

science is traditionally seen to have been established; second, the emergence of science as a professional activity in the early nineteenth century, when the social structures were established which provided the basis for the integration of science into the fabric of social life; and, third, the vast expansion *circa* 1900 of scientific education and research and the emergence of science-based technology and industry. These three phases of scientific development were echoed, indeed sometimes driven, by developments within Cambridge.

For the University as for the wider culture the 'scientific revolution' was embodied in the achievements of Isaac Newton. Elements of Newton's *Principia Mathematica*, which united terrestrial and celestial phenomena within a mathematical description of nature, provided the basis for teaching and examination in the Mathematical Tripos, established around 1750. Until the 1870s the Mathematical Tripos had primacy in shaping Cambridge science, but the eighteenth-century University provided few opportunities or incentives for the pursuit of science. The holding of one of the scientific professorships was frequently regarded as a sinecure, and scientific subjects did not form part of the curriculum; even in mathematics the traditional cast of Cambridge study hindered the assimilation of the work of the great continental mathematicians. The reforms of the Mathematical Tripos in the first half of the nineteenth-century – the introduction of analytic notation, the special focus on physical and geometrical topics (Cambridge's 'mixed mathematics') – had its result: George Gabriel Stokes and William Thomson, later Lord Kelvin (graduating in the 1840s) and James Clerk Maxwell (graduating in 1854) became the 'Cambridge' physicists who pursued physical theory as 'mixed mathematics', creating the 'Cambridge' school (as it has been termed) of mathematical physics. The industrial revolution generated new educational demands and expectations. At mid century a Royal Commission inquired into the University's capacity to adapt to 'the requirements of modern times'. It was in the 1870s that science in Cambridge began to take on a recognisably modern form; at the time, this was seen as marked by Maxwell's appointment as the first

Professor of Experimental Physics and the endowment of the Cavendish Laboratory. While this is the most famous example, there were contemporary developments in geology, physiology, botany, and zoology which led to the establishment of laboratories and museums, fostering a new scientific culture within the University. But only in the 1880s did the new Natural Sciences Tripos (established in 1851) begin to attract significant numbers of students; by 1900 it had become the most popular of the triposes. The University had entered the scientific age.

The authors of this collection of essays are drawn from within and beyond the University; they include historians of science and scientists currently active in Cambridge; and, whilst most essays are historical, some personal memoirs are included, offering a diversity of approach. Inevitably some important names have been omitted: Francis Bacon, Henry Cavendish, and John Herschel could all with justice have been included. Many other names could readily be added were it not for the demands of space and the need for chronological and thematic balance. The chronological balance reflects the increasing importance of science within Cambridge over the past 400 years and the changing relation between the University and 'Cambridge' scientists. Before 1800 such scientific associations were often tenuous; in the nineteenth century a 'Cambridge' scientist could very likely have limited professional contact with the University after graduation; but in the twentieth century, when scientific careers were more readily pursued within universities, close Cambridge associations – in education and in the pursuit of academic careers – have been judged appropriate in delimiting 'Cambridge' scientists. This book is a collection of scientific portraits, not a history of Cambridge science. But one feature requires special comment: only one woman is listed among the Cambridge scientists whose careers are discussed here. Women's education at Cambridge only began in the 1870s, and was for many years subject to severe and shameful impediments. As a result, women faced significant problems in undertaking laboratory work. As noted in several of the essays that follow, women featured significantly in

'Cambridge' science in the twentieth century and pursued distinguished careers both within and beyond the University. To take two famous examples, both crystallographers: Dorothy Hodgkin, later a Nobel laureate, began research in Bernal's Cambridge laboratory in the 1930s; and Rosalind Franklin, whose work on DNA was fundamental to the Crick–Watson model, studied at Newnham; but their scientific achievements came later and elsewhere. With this important caveat, it is hoped that the essays that follow will give a picture of the range of the scientific associations of the University and of the place of science in its culture and history.

Further reading

Elisabeth Leedham-Green. 1996. *A Concise History of the University of Cambridge*. Cambridge: Cambridge University Press.

Richard Mason (ed.). 1994. *Cambridge Minds*. Cambridge: Cambridge University Press.

Edward Shils and Carmen Blacker (eds). 1996. *Cambridge Women, Twelve Portraits*. Cambridge: Cambridge University Press.

Peter Harman is Professor of the History of Science at Lancaster University. In the 1970s he lectured at Cambridge, where he was also a Research Fellow of Clare Hall. He was Zeeman Professor of the History of Physics at the University of Amsterdam in 1995. His interest in Cambridge science derives from his work on James Clerk Maxwell, and he has edited a collection of essays, *Wranglers and Physicists. Studies on Cambridge Physics in the Nineteenth Century* (1985). His books include *Energy, Force, and Matter. The Conceptual Development of Nineteenth-Century Physics* (1982), *The Natural Philosophy of James Clerk Maxwell* (1998), and a three-volume edition of *The Scientific Letters and Papers of James Clerk Maxwell* (1990, 1995, the third volume will be published in 2002), all published by Cambridge University Press.

I William Gilbert
STEPHEN PUMFREY

> As for the causes of magnetic movements, referred to in the schools of philosophers to the four elements and to prime qualities, these we leave for roaches and moths to prey upon.
>
> Gilbert, *De Magnete*, Book II, Chapter 3.

The reputation of William Gilbert (1544–1603) as a great scientific mind traditionally rests on three foundations, all of which are evident in the only book he published, the seminal *De Magnete* [On the Loadstone] (London, 1600). First, he discovered that the Earth was a giant magnet and, in order to establish the fact, inaugurated the modern science of magnetism. Secondly, he rightly boasted that the method evident in *De Magnete* was experimental, a radical break with the more textual methods used by his scholastic contemporaries. Finally, he distinguished between magnetism and electricity, which had hitherto been paired as similar, occult attractive principles; he even coined the noun *electricitas*, which was rapidly Anglicised as 'electricity'. Gilbert has been made a hero as 'the first experimental scientist', and he would come first, chronologically, in many surveys of scientific minds, not just of Cambridge minds. In Cambridge, he is immortalised in the name of Gilbert Road, a development built on land belonging to his college, St John's. As a Cambridge schoolboy, I entered my primary school every day from Gilbert Road, regrettably ignorant of the existence of the eponymous scientific hero.

Nevertheless, Gilbert's inclusion in this collection is probably the most controversial. This is not because the extent of his fame in his lifetime was limited to be one of England's most eminent doctors, who rose to become President of the College of Physicians, and a royal physician to both Queen Elizabeth I and James I. There are three more profound reasons. First, like Sir Isaac Newton, he did not practice science.

As Peter Harman's introduction makes clear, our modern discipline came into existence two centuries later. Like Newton, Gilbert described himself as a natural philosopher, although Newton differed from him by emphasising the importance of mathematics and of clear methodological rules in the investigation of nature.

Secondly, some of Gilbert's central beliefs were decidedly pre-scientific. He held that the planets possessed some form of soul, the earth's being a magnetic one. He believed in divine cosmic harmonies, and he practised astrology. If these beliefs do not exclude Gilbert, we might consider the first Cambridge scientist to be Dr John Dee, graduate of St John's College in 1545, founding fellow of Trinity College, promotor of Euclidean geometry, and interrogator of angels.

Thirdly, like Francis Bacon, Gilbert's attitude to the academic values of Cambridge University, indeed of university natural philosophers everywhere, was hostile and dismissive – witness the quotation that begins this chapter. Gilbert would have agreed that his mind flourished only when he left the groves of academe for the cultural and economic dynamo of London, which in late Elizabethan times was the booming centre of an emerging imperial power.

Let us begin with a brief biographical portrait, and then focus on Gilbert's natural philosophical achievements, before concluding with his problematic relationship to Cambridge. He was born in Colchester, Essex in 1544, the eldest son of Jerome and Elizabeth. The Gilbert family came from merchants of relatively recent wealth, and Jerome benefited by gaining a university education and a profession – law. As the eldest son of middling pseudo-gentry, William was likewise prepared for a professional career, in the expanding field of medicine. He went up to St John's College in 1558 from Colchester Grammar School, and proceeded to a BA in 1561. He was admitted to a fellowship, and received his MA in 1564. To do so he probably lectured on Aristotle's physical works *De Caelo* and *Meteorologica*. He then studied for an MD, which was awarded in May 1569.

There are no signs that he was discontented with the academic world of Cambridge at this stage. Indeed he took on posts at St John's,

becoming mathematical examiner in 1565 and 1566, and bursar in 1570. His only surviving books come from his time at St John's and they are perfectly traditional: two volumes of Galen, one of Aristotle's natural philosophy, and Matthioli's *materia medica*. There is no truth in the story that, because he had a low opinion of the Cambridge medical faculty, he took a medical degree abroad, as did William Harvey and other ambitious physicians. Gilbert's glittering, home-grown medical career was matched step-by-step by Harvey's father-in-law, his friend and fellow Johnian, Lancelot Browne.

There is then a gap in his *curriculum vitae*, because records of Gilbert's life and work are lacking. He died of the plague, and his effects were probably burned. Other papers and instruments that he bequeathed to the College of Physicians perished when the College, like his London residence, was destroyed in the Great Fire of 1666. The best guess is that, like many young physicians, Gilbert moved to London in order to build up a medical practice. He succeeded, and was already a Censor in the London College of Physicians in 1581, putting him near the apex of its forty-odd Fellows.

To become a royal physician required not only the College's backing but also that of powerful nobles. Gilbert had the best. By 1581 he was already a client of Robert Dudley, Earl of Leicester, and later served the family of William Cecil, Lord Burghley, amongst others. These patrons probably influenced not only Gilbert's medical, but also his natural philosophical career, because Leicester and Burghley patronised networks of mathematical practitioners, such as John Dee and Thomas Digges, directing them to military and naval research in the service of the state. Thus, three months prior to the defeat of the Spanish Armada, Gilbert (and Browne) were named as 'fytt persons to be employed in the said Navye to have care of the helthe of the noble-men, gentlemen and others in that service'.

Through courtly contacts, Gilbert got to meet and admire famous mariners, such as Sir Francis Drake and Sir Thomas Cavendish (England's first circumnavigator), and leading theorists of navigation, such as William Barlowe and, most influentially, Edward Wright. From

these eminent Elizabethans, Gilbert learned about the importance of magnetic navigation using the compass, and became aware of the general lack of understanding of the compass and magnetism. Indeed, Edward Wright, once a fellow of Gonville and Caius but subsequently mathematician to the Earl of Cumberland, collaborated closely with Gilbert in the composition of *De Magnete*, Wright provided magnetic compass data, the latest navigational theories, and actually wrote parts of it.

Gilbert's book is infused with an empiricist rhetoric that preempted his younger courtier colleague Bacon. He insisted that those who worked with nature, like navigators, metallurgists, and farmers, understood more about the nature of the Earth, and earthly matter, than did professors of scholastic Aristotelian philosophy. It is, however, not plausible to assume that either a commitment to improve magnetic navigation, or an empiricist's determination to investigate the loadstone thoroughly, was sufficient motivation for Gilbert to devote (according to some sources) eighteen years and £5 000 in the preparation of a scientific work 'on the magnet'. That said, *De Magnete* was an unrivalled synthesis of past views (invariably criticised), reliable reports, and new experiments. Prior to *De Magnete*, the most exhaustive and empirical treatise had been *On the Loadstone*, which formed Book VII of the 1589 edition of Giambattista della Porta's *Natural Magic*. A flavour of Gilbert's experimentalism can be gained from his careful refutation of Porta's conclusion that an iron needle rubbed with diamond also points north.

> Now this is contrary to our magnetic rules; and hence we made the experiment ourselves with seventy-five diamonds [!] in the presence of many witnesses, employing a number of iron bars and pieces of wire, manipulating them with the greatest care while they floated in water, supported by corks; yet never was it granted to me to see the effect mentioned by Porta.

Recent historians have taken seriously Gilbert's cosmological beliefs as his prime motivation. His central philosophical dogma was

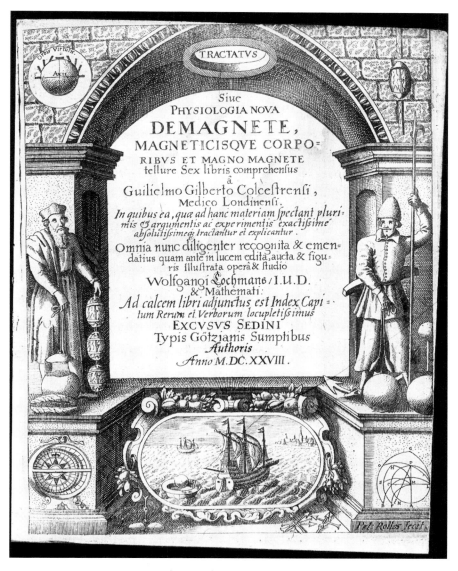

FIGURE 1.1 Title page of Dr William Gilbert's 1628 edition of De Magnete.
Source: The Whipple Library, University of Cambridge.

that the Earth was a noble part of the cosmos, seeming to possess animate powers of the kind ascribed to planets. Consequently, he was harshly dismissive of the Aristotelian natural philosophy of the earth that he had pursued at Cambridge. Aristotelian philosophers, often called Peripatetics, divided the cosmos into a perfect superlunary realm, where stars moved themselves in circles, and, below the Moon, a corruptible terrestrial world composed of the four elements. Elemental earth was held to possess the passive qualities of coldness and dryness, and was therefore inactive. It descended naturally to the central point of the universe, furthest from the heavens; some Peripatetics even described the resulting stationary sphere as *'faeces mundi'*.

Freed from the university constraints to uphold Aristotelianism, Gilbert argued vehemently that, despite increased mining and global exploration, '[t]he Aristotelian element, earth, nowhere is seen, and the Peripatetics are misled by their vain dreams about the elements'. Indeed, 'Aristotle's "simplest element", and that most vain terrestrial phantasm of the Peripatetics – formless, inert, cold, dry, simple matter, the substratum of all things, having no activity – never appeared to any one even in dreams'.

Quite why Gilbert rejected traditional matter theory might have been recoverable from his lost papers. The cold winds that sweep across the North Sea to Cambridge may have provided one reason: Gilbert remarked that it was typically narrow-minded of the Greeks to have classified elemental air as hot and wet! In general, we are forced to reconstruct an account from the six books that comprise *De Magnete*, and from the tracts assembled posthumously by his half-brother into a manuscript called *De Mundo* and presented to Henry, Price of Wales. This work, of which Bacon possessed a copy, was not published until 1651. Translated, its title is *A New Philosophy of our Sublunary World*, with the subtitle *A New Natural Philosophy in Opposition to Aristotle*. These are good indications of Gilbert's general project, and provide a wider context in which to read *De Magnete*.

Gilbert's project was not unique. He can be grouped with contemporary 'nature philosophers', such as Francesco Patrizzi and Giordano

Bruno (both of whom he cited and criticised), who developed new cosmologies influenced by Neoplatonism. Gilbert shared with Bruno a conviction that an Earth with planet-like powers would also exhibit the planet-like motions given to it by Copernicus in 1543. Indeed, Gilbert was one of only ten writers to have advocated a fully heliocentric cosmology by 1600. But, unlike Bruno and the others, he had little expertise in Copernican astronomy, and we cannot be sure whether his Copernicanism was a cause or a consequence of his matter theory. Gilbert's uniqueness, in both natural philosophy and cosmology, stems from his conviction that he had empirical proof of a new, anti-Aristotelian theory of active terrestrial matter. That proof came from his discovery of the Earth's magnetism, laid out in *De Magnete*.

Gilbert's evidence and reasoning exemplifies his unprecedented experimentalism, which impressed supporter and opponent alike, and which ensured that *De Magnete* was not ignored. There is, however, no coherent *method* beyond two working principles.. The first is his sceptical empiricism: his insistence that, since nearly all established explanatory concepts were wrong, one had to reason from securely observed phenomena. The second is what we can call his central principle of analogy. Gilbert argued that a model of the Earth, a *'terrella'* turned from natural loadstone, replicated all the magnetic phenomena of the Earth itself, such as the orientation of compass needles. With this principle Gilbert explicitly denied, as did Bacon, the Aristotelian doctrine that 'art' (technology) could not imitate nature. Therefore, the Earth could be experimentally investigated in the laboratory.

Gilbert's most significant experiments were conducted with miniature compass needles – he called these *versoria*, or 'rotation detectors' – which he moved over the surface of *terrellae*. Books II–V describe how Gilbert replicated four of the five 'magnetic motions' that he identified: coition, or the attraction of opposite poles; direction, or north–south alignment; variation, conceived of as a slight rotation away from true north or south; and inclination, or magnetic dip. Gilbert therefore concluded, by analogy, that the Earth itself was a giant spherical loadstone – a claim flagged in the full title of *De Magnete*.

One experimentally grounded (though erroneous) analogy interested both natural philosophers and navigators, for whom it offered an explanation of magnetic variation. Variation, or the angle between a compass bearing and true north, was the bane of navigators. By 1600 its reality was undisputed, although its complex pattern of distribution had yielded numerous theories. Some regarded it as an instrumental artefact, but learned English navigators like Edward Wright preferred Simon Stevin's 1599 hypothesis. For Stevin, variation was no artefact; it was distributed irregularly in geographically specific patterns. Recognising these patterns through compass observations offered navigators a limited way of finding longitude at sea, or a *Havenfinding Art*, as Wright entitled his English translation of the Dutchman's work.

Gilbert had a loadstone 'crumbled away at a part of its surface and so having a depression comparable to the Atlantic sea'. According to Gilbert, *versoria* moving over this imperfect sphere exhibited similar patterns of variation to those recorded by transatlantic mariners. Variation was thus the consequence of the Earth's geological deviations from perfect sphericity. The explanation not only confirmed Stevin's haven-finding method; it also allowed Gilbert to argue that the Earth was essentially a perfectly spherical magnet, whose magnetic poles were identical with its geographical poles. Such inferences prepared Gilbert for the climactic Book VI, which cannot be dismissed as a lapse into 'fuzzy medieval speculation', as one historian put it, if only because much of *De Mundo* elaborates upon it.

In Book VI Gilbert marshalled evidence that magnetism was the motive force of the Earth's Copernican motions. He may have been inspired by the thirteenth-century writer Petrus Perigrinus, who claimed that a spherical magnet suspended from its pole rotated every twenty-four hours. Gilbert typically tested the claim and rejected it, at least for ordinary magnets. But for the 'prime magnet', i.e. the Earth, Gilbert asserted that its soul-like magnetic power did indeed imbue it with a fifth magnetic motion, that of rotation. Magnetism both rotated the Earth diurnally and magnetically stabilised its axis of rotation. Gilbert cleverly evaded any clear statement about the Earth's annual

rotation, perhaps because he had no magnetic proof of it. In *De Mundo*, Gilbert went on to assert that each planet had its own specific power or virtue. The Earth's, and the Moon's, were magnetic; thus the Moon's orbit, and tides, were caused by magnetic attraction – an interesting adumbration of Newton's lunar theory. The sun had a luminous virtue, which 'predominated' and 'incited' the other planets to move around it. The virtues combined harmoniously to generate the planetary orbits. In this way Gilbert sketched out an experimentally grounded, natural philosophical dynamics for the Copernican system, the first to explain why a planet such as the Earth orbited the Sun, rotated stably on its axis in empty space, and exerted an attractive force on bodies in its vicinity.

His grand vision of a *philosophia magnetica* – a magnetic natural philosophy, not a science of magnets – accounts in large part for its appeal in the period prior to Newton's theory of gravitational attraction. As early as 1603, Johann Kepler wrote that he could 'demonstrate all the motions of the planets with these same [Gilbertian] principles'. He attempted to do so in his *Astronomia Nova* (1609), granting all the planets complex pairs of magnetic poles and calculating the resultant forces. Stevin promoted magnetic Copernicanism in the Dutch Republic. Galileo was another early Gilbertian, and the Inquisition criticised him for praising the 'perverse and quibbling heretic'. In 1657 Christopher Wren named Gilbert and Galileo as the two 'assertors of philosophical liberty'. Together with John Wilkins and Robert Hooke, Wren perpetuated Gilbert's model of attractive celestial forces into Newton's era.

With *De Magnete* popular among the seventeenth century's 'new philosophers', it is not surprising that Jesuit natural philosophers published more works on magnetism than did any other school of thought. Niccolo Cabeo paved the way in his *Philosophia Magnetica* of 1628, and brilliantly showed how Gilbert's discovery was, in fact, compatible with Aristotelian matter theory. Catholics troubled by the Galileo affair argued that the Creator had used magnetism as an additional cause of the Earth's immobility.

Since Gilbert's magnetic philosophy was closely tied to Copernicanism, it is superficially surprising that Gilbert did not attempt to discover any quantitative magnetic laws that could have advanced the emerging field of physical astronomy. In fact, just as Gilbert was the only non-astronomer amongst the early Copernicans, so was he unique in maintaining the conservative, scholastic distinction between mathematics and natural philosophy. Gilbert insisted that natural philosophers alone discovered physical causes, whilst mathematicians invented non-physical, fictional hypotheses to 'save the appearances' of the heavenly bodies. Gilbert was delighted that magnetic philosophy gave a real, physical, magnetic existence to the Earth's poles and parallels of latitude, entities that had previously been mere projections on to the Earth's surface of a revolving heavenly sphere. But, by the same token, Gilbert praised those astronomers who invented fictional orbits. Gilbert wrongly claimed that Copernicus and Tycho Brahe were fictionalists in this traditional sense. Indeed, Gilbert had an historical theory of cosmology, according to which error began in classical times when natural philosophers first misinterpreted mathematicians' orbits as real paths. Gilbert clearly shared Bacon's disregard of the power of mathematics to reform science.

Although Gilbert's position might seem backward looking, he had his reasons. They are evident in his concept of a magnet's 'sphere of virtue'. This *orbis virtutis* is only loosely related to later ideas of the magnetic field. Certainly Gilbert pointed to experimental proofs of magnetism's immateriality, for example that it passed through non-ferrous solids. Magnetism's immateriality was, for Gilbert, the important distinction between it and other traditionally occult attractions, such as electricity. (Gilbert's few electrical experiments were designed to show that 'electricity' was affected and, therefore, mediated by material effluvia, such as water vapour.) *De Magnete*'s diagrams are also reminiscent of modern 'lines of flux'. Gilbert was well aware that magnetic power decreased with distance and mobilised such demonstrable and law-like behaviour as further evidence that magnetism was no ordinary occult quality.

However, Gilbert ultimately denied that magnetic power could be analysed using mathematics, because mathematics was incapable of capturing its vitalist properties. Gilbert struggled for a language to describe the Earth's magnetism. The magnetic virtue in a loadstone was derivative of the whole Earth's more noble power. He shied away from a fully animistic model of this power, describing the Earth as 'as it were, ensouled' or as having a 'quasi-animate' power. Nevertheless, he considered that the Earth and other planets were able to respond to each other's powers. This resulted in a concerted heliocentric harmony that was irreducible to mathematical quantities. In modern terms, Gilbert held that the planets' mutual pertubations were too complex to analyse. It is an irony that the first plausible physicist of Copernican cosmology should have resisted Copernicus's own intention of uniting mathematics and physics. Gilbert's attitude was conventional, but there is another explanation. Edward Wright admitted to Mark Ridley, Gilbert's fellow physician, magnetician, and lodger, that Gilbert was 'not skilled in Copernicus' and needed instruction from one Joseph Jessop, another London physician and erstwhile fellow of King's. Gilbert seems to have concluded that mathematical difficulties represented mathematical impossibilities. The inability of Newtonian mechanics to solve the many body problems presented by planetary perturbations might be adduced in Gilbert's favour.

Gilbert's traditional subordination of mathematics to natural philosophy raises a problem in understanding *De Magnete*. One of the impressive, 'modern' features of *De Magnete* is its very use of mathematics, especially of practical techniques relating to navigation. Book VI concludes with two very technical chapters on Copernican models of the precession of the equinox. Book V contains instructions on how to make and use a magnetic inclinometer. There is also a complex, accurate geometrical nomograph that allowed sailors to read off their latitude from inclination measurements –another promising application of magnetic philosophy to navigation. Book IV, chapter XII contained state-of-the-art instructions for calculating variation from observations of bright stars. These practical elements, combined with

Gilbert's philosophising and Aristotle bashing, have influenced historians, especially Marxists, to read Gilbert as the first to effect a synthesis of practical or experimental expertise with philosophical rigour.

In fact, Edward Wright admitted to Ridley that he had written Book IV, chapter XII. I strongly suspect (as did Ridley) that Wright was also responsible for other technical sections. He certainly collaborated on the final stages of publication. Moreover, Wright's address to the reader presented *De Magnete* as primarily a contribution to magnetic navigation, and only secondarily as the creation of a magnetic philosophy.

This sheds interesting light on one of the great mysteries of Gilbert's work. Whilst *De Magnete* is thoroughly experimental, replete with geomagnetic data, bristling with new instruments, and full of practical applications, *De Mundo* is in a different genre. It is largely speculative Renaissance nature philosophy, resembling Patrizzi's discursive anti-Aristotelianism. It develops the speculative magnetic cosmology, and it adds, to the elemental theory of magnetic earth, a theory of aqueous and oily effluvia unsubstantiated by any experiment. Indeed, there are no new experiments in *De Mundo*; the empirical arguments draw on common-sense or anecdotal observations. It is tempting, therefore, to suggest that the rigorously experimental *De Magnete* is not the natural philosophical treatise Gilbert himself wanted to write, but a product of the collaboration with, and influence of, Wright.

The possibility that *De Magnete* arose out of the fusion of two Cambridge minds brings us back to Gilbert's debt to his Cambridge milieu. Obviously his Cambridge training in mathematics, natural philosophy, and medicine was crucial. At St John's, Gilbert acquired the professional medical skills that would propel him into the courtly and maritime communities of London. He also absorbed the traditional disciplinary boundaries of natural philosophy, the interconnection of matter theory and cosmology, and the (ir-)relevance of mathematics. Like all revolutionaries, Gilbert discarded much less traditional conceptual baggage than he thought. The opening chapters of

De Magnete, thorough, critical reviews of existing opinion, follow humanist dialectical method. The structure and chapter headings of *De Mundo*, such as '*De Aqua et Terra*', '*De motu gravium et levium*', '*De telluris loco*', and '*Meterorologia quid sit*', come straight from the scholastic curriculum, even if Gilbert denied that the entities existed or that the doctrines were right. It has been plausibly argued that his concept of *orbis virtutis* derives from the Aristotelian *sphaera activitatis*, and that his notion of the soul is Thomistic.

Moreover, Gilbert's scientific and medical careers both developed in the company of scholars who made up his Cambridge milieu. The community of mathematicians that flourished in Elizabethan Cambridge, and supplied London with lecturers in navigation, was literally instrumental in transforming magnetism into a topic for his experimental investigation.

Any innovative scientist, however, needs a disciplinary training and a community of intellectuals with whom to develop new ideas. In Gilbert's case we cannot point to any positive intellectual influence that he encountered through the university, as we can for Newton, who was influenced by the Cambridge Platonists, and for Harvey, whose anatomical discoveries depended upon the methodology he acquired at Padua from Fabricius. Gilbert certainly did not exclude Cambridge 'science' from his criticisms of Aristotelianism as dogmatic, stupid, stuck in the Renaissance cult of books and antique authorities, and shored up by long familiarity, proscriptions against free thought, and its incorporation into theology.

Of course, almost all Elizabethan natural philosophers were university educated, many at Cambridge. But in Gilbert's period the innovative action was in London, where noble patrons supported Paracelsian physican–philosophers like Thomas Moffett and Robert Fludd, or mathematicians like Digges and Wright. After 1596 London also had trusteeship of the foundation of the merchant Sir Thomas Gresham. Gresham College was designed to remedy Oxbridge's lack of relevance to the commercial world. Henry Briggs, another Johnian, moved to the metropolis, was appointed the first Gresham Professor of

Geometry, and collaborated with Gilbert and Wright in applying the discovery of the dip–latitude relation. The London milieu also shaped Francis Bacon's ideology, and introduced him to projects and examples of progressive technology that he compared favourably against conservative university philosophy.

Of course, the curriculum followed, and rejected, by Gilbert and Bacon had once been progressive and vocational; it was designed to produce men of letters to fill clerical, legal and other positions in an expanding state bureaucracy. Bacon acknowledged its continuing utility in some of these areas. But Gilbert and Bacon heralded a new era of *philosophia naturalis plus ultra*, that looked beyond the limits of classical contemplative knowledge to a new, applied science. Do not universities still have a tendency to defend as scholarship the vocational learning of a previous era?

Further reading

William Gilbert. 1958. *De Magnete*, trans. P. Fleury Mottelay, 1893 reprint edn. New York.
Duane, H.D. Roller. 1959. *The 'De Magnete' of William Gilbert*. Amsterdam.
Sister Suzanne Kelly. 1965. *The 'De Mundo' of William Gilbert'*. Amsterdam.
Stephen Pumfrey. 1989. Magnetical philosophy and astronomy, 1600–1650. In *The General History of Astronomy*, eds. R. Taton and C. Wilson, Vol. 2., part A, pp.45–53. Cambridge.

William Gilbert (1544–1603): English physician and natural philosopher: discoverer of terrestrial magnetism, pioneer of experimentalism and early developer of Copernicanism.

Gilbert studied medicine at St John's College, and rose to become a royal physician. His book *De Magnete* [On the Loadstone, 1600] is a classic of emerging experimental science and was closely studied throughout Europe. In it Gilbert investigated magnetic phenomena exhaustively, establishing magnetism's immaterial nature and distinguishing it from electricity. His major innovation was to use a spherical loadstone or '*terrella*' as a laboratory model of the Earth. He thereby demonstrated the Earth's magnetism, and developed numerous laws and insights to govern the use of compasses in navigation. Whilst these practical applications of magnetism ensured many followers, his biggest impact was in

natural philosophy. Gilbert argued that the Earth's magnetic force was incompatible with Aristotelian science, and provided experimental proof of the Earth's motion. This 'magnetic philosophy' inspired Kepler, Galileo and Descartes, and provoked conservative responses from Jesuit scientists.

Stephen Pumfrey is Senior Lecturer in History of Science at the University of Lancaster. His 1987 doctoral dissertation was on William Gilbert and magnetic philosophy. He was a co-editor of and contributor to *Science, Culture and Popular Belief in Renaissance Europe* (Manchester, 1991), and has published numerous articles and chapters on the scientific revolution. He has close connections with Cambridge, having been born and educated there, including four years at Gilbert's College, St John's. His current research projects include a critical edition and translation of Gilbert's *De Mundo Nostro Sublunaria*, and a study of science and patronage in England, 1570–1626.

2 William Harvey

ANDREW CUNNINGHAM

William Harvey was a student at Gonville and Caius College in Cambridge, entering in May 1593 at the age of fifteen and staying until 1599, when he was twenty-one. Harvey is the celebrated discoverer of the circulation of the blood. This was, and still is, simply the most important discovery about the anatomy and physiology of the human and animal body that has ever been made. All of our modern physiological understanding is based on this discovery. Harvey made the discovery on his own, in the course of his private researches in London sometime around 1618, and he was so overwhelmed by what he had discovered that it was not until ten years later, after repeatedly presenting it to the criticisms of his colleagues in the College of Physicians in London, that he could bring himself to publish it.

Harvey was born on 1 April 1578 in Folkestone in Kent, where his father was mayor on four occasions. Harvey was educated at The King's School, Canterbury, and then he was awarded the Matthew Parker scholarship, which was restricted to boys from The King's School, and this took him to Gonville and Caius College in Cambridge. This college had been founded out of Gonville Hall, a somewhat decayed hall for students, by John Caius (pronounced 'Keys'), a celebrated physician, in 1557. At that time Caius (as the college is known for short) was the most medical of all the colleges of Cambridge or Oxford, with more people studying medicine there than at any other college. It has continued to have a long and very distinguished tradition of teaching and research in medicine and the medical sciences right up to today. Matthew Parker, Archbishop of Canterbury from 1559 to 1575, had been a friend of John Caius, and it is perhaps through this relationship that Parker decided to found this medical scholarship – the first in England – at Caius College. William Harvey was the fourth student to

hold it. The chance of being sent to the right school, and being able to take this scholarship to Cambridge, thus determined Harvey's career as a doctor.

This scholarship lasted for six years, and it would have been possible under the university statutes for Harvey to have spent the whole six years pursuing the Bachelor degree in medicine, without having to take an arts degree first. However, while Harvey did hold this scholarship for almost the full six years, he at some point decided to take the Bachelor of Arts degree in 1597. This suggests that he was primarily studying the conventional philosophy course, rather than medicine, at least for the first three years. He did not then proceed to take the Bachelors degree in medicine at the end of the sixth year of his scholarship. So it is not at all clear how much medicine Harvey studied at Cambridge – and hence whether Cambridge as his *alma mater* deserves any credit for training Harvey as the great anatomical researcher he turned out to be.

In Harvey's student days Cambridge was not a particularly good place to study medicine, though it was improving. Cambridge was a university where the masters had for centuries been dominant in the organisation of teaching, and it was the regent masters or regent doctors, as they were called – those who had relatively recently graduated in each discipline – who were expected to do most of the teaching of the subject. This teaching consisted of the reading of authoritative texts out loud, commenting on them, and running disputations on them for the junior members of the faculty. All this was conducted in Latin. However assiduously it may be performed, such teaching leaves few traces. For medicine the texts being read out loud, commented and disputed on, would have been primarily those of two ancient Greek physicians: the thirty or so anonymous works which go under the name of Hippocrates, and the very extensive writings of Galen, which had been recently recovered, translated from Greek into Latin, and established as the basis of medical knowledge in the universities. Foremost among these works of Galen were his anatomical texts, and people had come to recognise in the course of the sixteenth century

that Galen possessed a formidable knowledge of human anatomy, which he had acquired from the dissection of animals, especially apes. The revival of this anatomical knowledge was perhaps the most important, and certainly the most spectacular, of the developments in medicine in the universities of Europe in the sixteenth century. But Cambridge, like Oxford, was far from the centre of such activities. What we see here in England in Harvey's day are two relatively insignificant universities in a marginal country on the edge of European intellectual life, but where attempts were being made to bring them up to the standard of the best universities abroad, such as those of northern Italy and of France. One of the steps which had been made to make Cambridge more like an Italian or French university was the introduction of the professorial system and of paid lectureships. By Harvey's student days this meant that there was one Professor of Medicine at Cambridge, appointed by the crown. In addition to teaching Hippocrates and Galen, this Professor was expected to perform anatomies, as would have been expected of such a professor on the continent, but nothing was done to secure bodies for him to dissect. A private benefactor, Thomas Linacre, a physician who had been at university in Italy and then had founded the College of Physicians in London, had endowed two lectureships in medicine, one at Cambridge and one at Oxford, and the Cambridge one was located in St John's College. This lecturer had to read Galen's books on the maintenance of health, on the method of healing, on the properties of foods, and on simples (that is, on plants used as medicines) 'to everyone wishing to hear'.

Such continental innovations had been further developed by John Caius in the college he refounded, and at which Harvey was a student. Caius himself had studied medicine at Padua, in northern Italy, and while there he had participated in translating some of Galen's works, and had witnessed the re-establishing of anatomy as the basis of Galenic medicine. Caius certainly did what he could to promote the study of the latest – which was of course also the oldest, the Greek – form of medicine in his college. He not only provided for medical fellowships, and gave directions for the method of study to be engaged in

by the medical Fellows, but he also acquired from the crown the right to the bodies of two executed malefactors for the fellows of Caius College to dissect, twice a year, so that the medical fellows and students could learn anatomy. Twenty-six shillings and eight pence were to be spent every winter by those studying medicine in the college to carry out the dissection and bury the body afterwards with all due reverence. So, whether he saw anatomies performed there or not (which we do not know), Harvey's own college had the best opportunities for witnessing human dissection of anywhere in England. In this sense Caius College deserves the compliment that Sir Charles Scarburgh gave it in a celebratory speech after Harvey's death. Scarburgh said that at Caius, Harvey 'drank in philosophy and medicine from the purest and richest spring of all, if there be such another dedicated to Apollo in the British Isles'.

So we can say that, at Caius College, Harvey was at the best possible place for learning academic medicine in the whole of England. However, his aspirations were higher than this. For in early 1600 Harvey went to the best university in the world for learning medicine, the University of Padua. Six years earlier, Shakespeare, through the mouth of Lucentio in *The Taming of the Shrew*, called it 'fair Padua, nursery of arts', and of all the arts Padua was the greatest nursery of medicine in its day. It was here that William Harvey developed his passion for anatomical research, which was, all unknown to him, to lead him to discover the circulation of the blood. Just as he had been steered to Cambridge and Caius College by the chance of him first attending The King's School in Canterbury, so now, as a student at Caius College, he was assisted in his choice of Padua both by the example of John Caius himself, and also by encouragement in the college statutes for the medical fellows to study abroad for three years at Padua, Bologna, Montpellier, or Paris.

Unlike Cambridge, Padua University was run by politicians. But they were the representatives of the greatest trading power of the day, the Republic of Venice. They ran it as a business. The *Riformatori dello studio*, the committee which controlled the university on the behalf of

the Senate, sought out the best professors in the world as teachers, and made it possible for non-Catholics to attend as students. The *Riformatori* were rewarded by a constant stream of student visitors from all over Europe, bringing their money with them to make Padua flourish, seeking to study with the best professors the philosophy (or arts) subjects and medicine, the two areas in which the university specialised.

Shortly after arriving in Padua, Harvey was elected Consilarius for the English Nation, which meant that he acted as representative of part of the student body. He received the degree of MD in April 1602. This is all we know directly about his time in Padua. But we certainly know, from the research that Harvey engaged in for the rest of his life, that the most important part of the teaching for him at Padua was the anatomical course run by Professor Fabricius ab Aquapendente (Girolamo Fabrizie). Fabricius took the ancient Greek philosopher Aristotle – not the Greek physician Galen – as his model in anatomising, and this meant that he studied the anatomy of the animal body, not just the human body, which was very unusual. When Harvey heard him, Fabricius was researching and lecturing in a new anatomical theatre (which still exists), and he was also writing a book he called *A Theatre of the Whole Animal Fabric*.

Harvey returned to England in 1602 or 1603, and settled in London where he established a private medical practice. He married Elizabeth Browne, daughter of Dr Lancelot Browne, in 1604. From 1609 he was physician at St Bartholomew's Hospital, for which he was paid £25 a year, and for which he had to attend one day a week to care for the poor patients of the hospital. He retained this position for thirty-five years. As a London practitioner, Harvey had to join the College of Physicians in order to have permission to practise; he was given licence to practise in 1603, made a Licentiate of the College in 1604, and a Fellow in 1607. Harvey was a very committed member of the College, regularly attending the meetings at which the College proceeded against unlawful practitioners. He held all the offices of the College (Censor 1613 etc., Elect 1627, Treasurer 1628) except that of President,

which he declined when approached in 1654. He undertook the Lumleian Lectures for the College from 1615 for forty years. These lectures were given to the apprentices of the surgeons, twice a week on a six-year cycle, with a five-day dissection of a specified region of the human body every winter. His Latin lecture notes for this series, with all the additions he made over the years, survive in the British Library. In his later years, when he was famous throughout Europe, Harvey encouraged the other Fellows of the College of Physicians to undertake anatomical research together and to publish their findings, even leaving money in his will for an annual feast and 'an exhortation to the Fellows and members of the College to search and study out the secret of Nature by way of Experiment'. He donated a new building to the College as a library and depository for rarities and the basic ingredients of medicines in 1651–2, but this was burnt down in the Great Fire of London. It is clear that the College of Physicians played the same role of learned club for him that the colleges of Cambridge today still play for their fellows.

William Harvey took up private research on anatomy immediately on his return to England. He was, John Aubrey wrote, 'the first that I hear of that was curious in Anatomy in England'. He researched primarily at home, and over a period of fifty years he vivisected and dissected all the animals that came to hand. His own list included hens, geese, pigeons, ducks, fishes, shell-fish, molluscs, frogs, snakes, bees, wasps, butterflies, silkworms, sheep, goats, dogs, cats, and cattle. He also dissected 'the most perfect of all creatures, man himself', both the corpses of those who died at the hospital and, for his anatomical teaching to the students, the corpses of those who had been put to death by the state for crimes. His inquisitiveness was such that he even dissected his wife's pet parrot after it had died, and even attended the autopsy of his own father (though he may not have dissected him).

Harvey's great discovery was that the blood circulates in all animals, including humans. He made the discovery around 1618, and he published on it in 1628. The immediate impact of Harvey's discovery was as great and as controversial in its time as Charles Darwin's

theory of evolution was to be in the nineteenth century. The reason for this is that it seemed to contradict everything that had been taken for granted for fifteen hundred years about the functioning of the body and the movement of the blood. Before he put forward his discovery, the account of the workings of the body that was the basis of medicine was that which had been established by Galen, the Greek physician who lived in Rome in the second century AD. This was what all medical students were taught, and what was understood too in popular culture, as for instance in the plays of Shakespeare. Galen had shown that there are *two* systems of the blood, almost completely separate from each other. The *veins* carry blood which has been made in the liver from the food, and this spreads to all the parts of the body, and each part of the body attracts from it what it needs for its nourishment. The system of veins is based on the liver. By contrast, Galen claimed, the *arteries* carry a very refined portion of blood, imbued with 'vital spirit', drawn in from the air as we breathe, and imparted to the arterial blood in the lung. This blood serves to 'vivify' or give life to the parts of body. The 'vital spirit' is consumed by the parts of the body, together with some of the arterial blood. In this system just a little blood is needed to pass through from the venous system to the arterial, to keep it topped up. Physicians and anatomists had therefore convinced themselves over the centuries that they could detect tiny pores in the wall (*septum*) between the right ventricle of the heart (venous blood) and the left (arterial blood), through which this small amount of blood passed. So everyone knew that the blood moved, but they knew that it was two different kinds of blood, which moved in two separate systems for two separate purposes. In both systems of blood vessel, the blood moved outwards and was consumed in the body: it did not return to the heart or 'circulate' around the body. The whole of medical understanding was built on this doctrine. And no-one was looking for a new route for the blood, or for the two systems of the blood to be seen as a single system. After all, Galen's account explained why the blood in the veins (the food for the body) looks darker than that in the arteries (lightened by 'vital spirit'), why the arteries pulsate (presence of 'vital spirit')

while the veins do not, why the walls of the arteries are thicker than those of the veins (to contain the lively 'vital spirit'), why blood spurts from an artery more than from a vein (again, the liveliness of the 'vital spirit'), and even why you die quickly if you puncture an artery, yet not if you puncture a vein. The whole system made anatomical sense, and was also highly common-sensical.

In this context, we need to ask what strange enquiry Harvey must have been engaged on that led him to see things completely differently. It was not, as some older historians used to maintain, simply a matter of Harvey using his eyes better than other researchers, nor the fact that he used experiment, for his opponents and rivals did this too. Indeed, with respect to using one's eyes, the circulation of the blood is not something which is visible: it is the *deduction* of an argument about anatomical pathways, the rate of pulsation, the capacity of the chambers of the heart, and so on. The circulation of the blood is certainly *not* something lying in front of one's eyes, just waiting to be found!

So what was Harvey's particular question, or his research problem? It turns out that this is where his time at Padua University was so crucial. Because what Harvey does is simply take up and continue the anatomical programme of his teacher there, Professor Fabricius. This is something he could not have learnt at Cambridge, since Fabricius was the only practitioner in the world of this particular approach to anatomising. But his time at Cambridge did prepare him in a particular way for this anatomising programme, though admittedly he could have received this preparation at any university in Europe. For Harvey had studied the arts or philosophy course at Cambridge and this, like the arts course in all other universities, was built around the works of the ancient Greek philosopher Aristotle. Now, it was Aristotle's programme of anatomising that Fabricius had been reviving. Almost no-one else could see in Aristotle's animal books a programme of anatomising to follow, but Fabricius could. It was this that Harvey copied and continued. So Harvey's exposure as an undergraduate at Cambridge to Aristotle's books on the soul, on analytic logic, on physics, on metaphysics, all came in very useful to his understanding

of how to undertake research, especially in following Aristotle's approach to investigating the anatomy of animals.

Fabricius had already been Professor of Anatomy for thirty-five years when Harvey arrived in 1600, and had spent these years reviving the anatomical programme of Aristotle and studying the same kind of things as Aristotle, such as the generation of animals, respiration, the local motion of animals – and all not on man alone, but on animals or, more properly put, on 'the animal'. In the course of this work Fabricius had discovered the so-called 'valves' in the veins. He thought they were there to slow the blood down as it moved out from the heart through the veins, bringing nutrition to all the parts of the body. So when Harvey took up anatomy on his own account in London, he too studied the generation of animals, respiration, and the local motion of animals.

The particular private research project that Harvey was engaged on, in the course of which he discovered the circulation of the blood, was on the heart, its vessels (i.e., the arteries, for it was believed that the veins come not from the heart but from the liver), and on the motion of the heart and arteries. For Aristotle, the heart was the centre of what was called 'the vegetative soul', and for Harvey too the heart of an animal was the most important organ. He called it 'the foundation of life, the prince of each animal, the sun of the microcosm'. Because he was investigating the movement of the heart and arteries in *all* animals, and not man alone, Harvey was not concerned with the lung, since only some animals have a lung, and this made his research project different from that of anyone else. He was using both warm- and cold-blooded animals, including fish and snakes, in his experiments and thus was able to observe the details of the heart's movement by seeing it beating slowly in cold-blooded and dying animals. When, after several years of research, Harvey had worked out the capacity of the chambers of the heart, the frequency of the heartbeat, the competence of the valves at the entrances and exits to the chambers of the heart, the competence of the flaps in the veins that his teacher Fabricius had discovered (and hence recognised them as being valves), and when he had also taken into account the relative size of the blood

vessels, he was forced to conclude (he says) that the blood must go 'as it were in a circle' (*quasi in circulo*). But it was a conclusion he came to with much hesitation, writing that 'those things . . . are so new and unheard of, that not only do I fear harm to myself from other people's ill-will, but likewise I fear that every man will be my enemy, so much does custom and doctrine once received and deeply rooted prevail with everyone'.

Why was Harvey shocked by this discovery? He had come to this conclusion because he felt obliged to accept the evidence of the experiments and his own eyes. This conclusion (viz. that the blood must circulate) explained why the parts of the heart and arteries had the shape, movement, and function that they had. He had found out the 'final cause', as the Aristotelians called it, of these parts: the reason why they are how they are and have the movements or actions that they have. But he could not find out the 'final cause' of the circulation of the blood itself, he could not find out *why* it circulated. As an Aristotelian therefore, he had failed to answer this most important of questions, and he was very aware that his account of this discovery was incomplete in this respect. This is probably why he sought the confirmation of his fellow members of the College of Physicians before he dared publish on it. He wrote to them:

> Since this only Book does affirm the blood to pass forth and return through unwonted tracts contrary to the received way, through so many ages of years insisted upon, and evidenced by innumerable, and those most famous and learned men, I was greatly afraid to suffer this little Book [to be published] . . . if I had not first propounded it to you, confirm'd it by ocular testimony, answer'd your doubts and objections, and gotten the Presidents verdict in my favor.

The distinctive features of Harvey's account of the movement of the heart and arteries are that *all* the blood in the body is pumped around the body rapidly and continuously by the force of the heart, outward from the heart through the arteries, and *returned* to the heart through the veins. In this Harveian understanding of circulation, all

the blood constitutes *one system*, not two, and the two types of blood vessel, the arteries and the veins, are parts of *one blood transport system*, and not two. Moreover, this concept of *unidirectional* circulation is built on a particular understanding of the functions of the flaps in the heart and in the veins, i.e. that they are *valves* and that they are competent in their functioning. All these features together are definitive of Harvey's concept of the circulation, and every one of them was new with Harvey.

The reason that most other people were shocked by this discovery was different from Harvey's own reaction to it. The horror of other physicians at Harvey's discovery is understandable. If Harvey was right – if the blood was in only one system and it was pushed around the body constantly by the action of the heart, from the arteries to the veins – then the whole of medicine was threatened. In particular, all the detailed logic of bleeding as a cure (where you bleed, how much, how frequently, whether away from or towards a diseased part), a logic built up over many centuries, was rendered useless. This was the basis of the objections of James Primrose, an English doctor who wrote against Harvey in 1630. In France, Pierre Gassendi objected to Harvey's denial of the pores in the septum. Some objectors could also offer experimental evidence which seemed to point against the existence of the circulation, or alternative explanations of Harvey's experiments. In general, those who opposed Harvey were concerned exclusively with the consequences of the circulation for man, whereas Harvey himself always studied anatomical phenomena in all animals.

Harvey suffered personally as a result of publishing his account of the discovery. Not only did he lose patients from his medical practice, but the common people thought he was mad ('crack-brained'), all the physicians were against him and many wrote books against him. According to John Aubrey, the gossipy antiquarian, it was only 'with much ado at last, in about 20 or 30 years time' that the circulation of the blood 'was received in all the Universities in the world, and he is the only man, perhaps, that ever lived to see his own Doctrine established in his life-time'.

Although his discovery of the circulation of the blood was so controversial, William Harvey was no radical or controversialist. Quite the opposite. In his political as in his philosophical beliefs, Harvey was very conservative. He was and remained a Royalist throughout the long period of the Puritan revolution (from the 1620s) and the civil war (1642–9). He was a physician to the King (physician extraordinary to James I from before 1618, and then to Charles I from his accession in 1625), and he even accompanied King Charles I on the battlefield in 1642. He remained a Royalist after the beheading of the King in 1649 and during the Commonwealth or Interregnum period (1649–60). Unfortunately for him, he died before the monarchy was restored in 1660. With respect to his philosophical opinions, Harvey was equally old-fashioned. He was, as we have seen, a devoted follower of the ancient Greek philosopher Aristotle, because he found that following Aristotle gave one a highly fruitful approach to the investigation and understanding of created nature, calling him 'the most diligent investigator of Nature'. He apparently called the modern philosophers 'shit-breeches' – an expression which likens them to little children who defecate in their clothes. 'Go to the fountain-head', he advised someone wanting to study medicine, 'read Aristotle, Cicero, Avicenna'. Nor did he care for the new discipline of chemistry.

One Frenchman embraced the doctrine of the circulation from the moment he first heard of it, in a way which was to be particularly influential. This was René Descartes. He did not so much adopt the circulation as Harvey taught it, as co-opt it for his own purposes. For, even as he adopted the doctrine, Descartes changed it. In his *Discourse on Method* (1637), Descartes wanted to present the body as acting like a machine. One of the ways in which he did this was by showing the beating of the heart as an automatic system, sending out to the body blood which itself acted in an automatic way in accomplishing all the functions of the body. It is here that he adopted, and transformed, Harvey's new doctrine. Descartes hypothesised that the heat of the heart causes the blood in the heart to instantly ferment and expand, and that this in turn causes the heart to beat and the blood to be pushed

out into the arteries. Further, Descartes wanted the blood to consist of particles of many different shapes, so that they could passively filtrate through sub-visible sieves which he believed must exist in the glands. If, as he proposed, all and only the particles of one particular shape pass through a particular sieve, then, having passed through, these particles now constitute a physiological fluid with new properties, such as tears, saliva, sweat, milk, and so on. In this way Descartes hoped to offer an account of how these operations take place without special 'faculties' in the organs, and without the presence of soul overseeing them. Hence, he believed, he could show that the body acted automatically, and that this automatic action arose necessarily from the mere shape and arrangement of their parts. Harvey rejected both these explanations put forward by Descartes. Fermentation could not account for the expansion of the chambers of the heart because, he said, 'nothing in fermentation or bubbling up rises and falls as if in the winking of an eye', as Descartes's theory would require to happen to the blood in the heart. Similarly, Harvey could not be persuaded that the mere passing of a fluid through some imagined sieve could change the nature of the fluid: 'for it does not seem likely that a fluid, by a simple and sudden filtration, takes on another nature'. For Harvey, Descartes was an unwanted ally – and probably a 'shit-breech'.

While working on the heart and arteries, Harvey also performed extensive research on the generation of animals. His interest in this topic can be traced again back to his student days since it was a research area of his own teacher Fabricius, who was in turn following Aristotle. Like them, Harvey was also investigating generation in *all* animals, and his primary research animal was the chicken and the egg. His book on this research, On the Generation of Animals, was published in Latin in 1651. For Harvey investigation of the generation of animals was, like his work on the heart and its vessels, an investigation into the 'vegetative soul', that power or set of powers which controls all the vegetative operations of the animal body: all the operations that the animal body has in common with plants (i.e. everything except for sensing and thinking). How does the 'first matter' get shaped by the

'first efficient' cause? How can we find out about the vegetative soul from investigating the generation of animals? He expressed his goals for this project like this:

> We therefore (in accordance with the Method set out by us) will expound firstly in the egg, then also in other conceptions of different creatures, what is constituted first and what afterwards by the great divine power of Nature with foresight, inimitable intellect and admirable order; and then we will discuss what we shall have observed about the first matter from which, the first efficient by which, and about the sequence of generation and its economy; so that thence we might grasp something for certain about each faculty of the formative and vegetative soul from its operations; [and] of the nature of the [vegetative] soul from the members or organs [of the body] and from their functions.

Harvey confessed to having found as many new problems as he had solved, but his major finding was that all generation arose from an egg of some sort: there was no spontaneous generation of any living creature. Harvey also worked on other Aristotelian–Fabrician anatomical projects, such as the local movement of animals, but these researches were not published in his lifetime.

What then did Harvey owe to his time at Cambridge for his astonishing and ground-breaking discovery of the circulation of the blood? He never mentioned his time at Cambridge in any of his writings, and it is very difficult to find any direct effect of this time on his subsequent researches. For these, Padua was exclusively responsible. However, by the chances of his education – first The King's School, and then Caius College – Harvey was steered towards the priceless innovative teaching of Fabricius in Padua.

Further reading

Geoffrey Keynes. 1978. *The Life of William Harvey*, 2nd edn. Oxford: Clarendon Press.

Andrew Cunningham. 1997. *The Anatomical Renaissance: The Resurrection of the Anatomical Projects of the Ancients*. Aldershot: Scolar Press.

Robert G. Frank, Jr. 1980. *Harvey and the Oxford Physiologists*. Berkeley, CA: University of California Press.

William Harvey. 1653. *The Anatomical Exercises: De Motu Cordis and De Circulatione Sanguinis in English Translation*. London. Reprinted in an edition edited by Geoffrey Keynes, New York, Dover Publications, 1995.

Roger French. 1994. *William Harvey's Natural Philosophy*. Cambridge: Cambridge University Press.

William Harvey (1578–1637) discovered the circulation of the blood. After Cambridge he went to the greatest medical school of the time, at Padua, and studied there in 1600. Back in London from 1602, Harvey was soon successful and was physician to James I (and later to Charles I), but his main interest was in research. By 1618 he had a clear idea of the circulation, but he continued to experiment on this. He did not publish his results until 1628 in the poorly printed, slim book *Exercitatio anatomica de motu cordis et sanguinis in animalibus* (Anatomical Exercises on the Motion of the Heart and Blood in Animals), usually known as *De motu cordis*, now one of the great scientific classics. Harvey's work was as fundamental as Newton's work on the solar system. Another area of his work was generation; his book *On the Generation of Animals* (1651) describes his work on this, which was soon superseded by microscopic studies. His work on animal locomotion was not found until 1959.

Andrew Cunningham is currently a Wellcome Trust Senior Research Fellow in the History of Medicine in the Department of History and Philosophy of Science at Cambridge University. He is the author of *The Anatomical Renaissance: The Resurrection of the Anatomical Projects of the Ancients*, published by Scolar Press in 1997, and co-author with Ole Peter Grell of *The Four Horsemen of the Apocalypse: Religion, War, Famine and Death in Reformation Europe*, published by CUP in 2000.

3 Isaac Newton: Creator of the Cambridge scientific tradition
RUPERT HALL

In the mid seventeenth century, mathematics and science were accorded no greater importance in the University of Cambridge than in other universities throughout Europe. One hundred years later the position was quite different. Though the traditional academic 'exercises' still took place, the ability of graduands was judged by their performance in the Senate House Examination or Mathematical 'Tripos'. During the seventeenth century, traditions of teaching mathematical subjects, 'natural philosophy' (i.e., physical science), and medicine were modernised in many European countries, including Britain, but the influence of Isaac Newton (1642–1727) brought about particularly swift and far-reaching changes at Cambridge, his own university.

As Lucasian Professor of Mathematics for nearly thirty years from 1669, Newton set some of his own discoveries before his auditors (few enough) without ever proposing any general reform of education, while in private – in documents long unread and showing little desire to alter the balance between humane and mathematical or scientific studies – he increased the latter's importance. Most interesting in these drafts is the new role of a mathematically based natural philosophy, for which students were to be prepared by courses in geometry and mechanics, that is, 'the demonstrative doctrine of motions . . . For without a judgement in these things a man can have none in [natural] philosophy.' The latter subject Newton explained as the investigation of those matters which he himself had so far advanced in his *Mathematical Principles of Natural Philosophy* [*Principia*] (1687), beginning with an understanding of time, space, body, and motion, moving on to rational and fluid mechanics, astronomy and cosmology, then ending 'if the [lecturer] have skill therein' with knowledge of minerals, vegetables, and anatomy.

FIGURE 3.1 An allegorical monument to Sir Isaac Newton and his theories on prisms. Line engraving by L. Desplaces after D.M. Fratta after J.B. Pittoni, D. Valeriani and G. Valeriani.
Source: Wellcome Institute Library, London.

If Newton's ideas 'of educating youth in the universities', by remaining hidden, were without effect, the processes of academic development in Cambridge ensured that some approximation to them was realised after his death. By about 1750 the Mathematical Tripos, from earlier informal beginnings, was determining the merits of graduands; unlike all traditional university examinations for degrees it was a written test, and though the more distinguished candidates were called 'Wranglers' there was no formal disputation. From the beginning 'honours' were awarded to the successful candidates, who were listed in order of merit. Great prestige attached to a high place among the Wranglers or (from 1768 onwards) the award of a Smith's Prize, and a corresponding prestige was attached to the subjects in which candidates were examined in the Senate House. As with Newton's *Principles* and *Opticks*, on which so much of the work was then based, questions were by no means restricted to pure mathematics – rather they embraced much of the theoretical science of the day, especially astronomy, mechanics, and optics. This became Cambridge's 'mixed mathematics'.

Obviously the development of this academic pattern, unparalleled outside Cambridge, was only possible because there were senior men in the university who were highly competent in the mathematical sciences and especially in the innovations introduced by Isaac Newton in his two great works. Some were also skilled practical astronomers. Since 1707, Plumian Professors of Astronomy had buttressed the teaching of Newton's successors in the Lucasian Chair of Mathematics; two Colleges – St. John's and Trinity – established small observatories. There were also, from 1783, Jacksonian Professorships of Natural Philosophy; the holder was required to deliver in each year thirty-six lectures and to demonstrate thirty experiments. Meanwhile a quite distinct development had brought about the appointment of Professors of Chemistry (the first appointed in Newton's time). However, the greater responsibility for the development of the pupils rested with college teachers, who 'read' to their pupils the opening sections of Newton's *Principles* at least, his *Opticks*, and various other

books giving them a thorough training in mathematical techniques and the solutions to problems in the mathematical sciences.

One such College teacher was Thomas Rutherforth (1712–1771) of St John's, elected a Fellow of the Royal Society in 1743, who published *Ordo institutionum physicarum*. ('The order of instruction in natural philosophy', 1743). This slim volume lists more than a thousand propositions providing an introduction to the sciences of mechanics, hydrostatics with hydraulics, pneumatics, optics, and astronomy. For propositions that were not definitional, the author provided a reference to one or more books in which the topic was treated; sometimes the reference is to pages in Newton's own writings, at others to those of his followers from Keill and 'sGravesande onwards. The bibliography of these writers includes thirty-one titles. Clearly, anyone mastering all these works would have a thorough knowledge of Newtonian science, not to be acquired without much effort.

No continental mathematics was taught at Cambridge. The pure tradition of Newton's calculus of fluxions was maintained. The Leibnizian calculus used to great effect by so many outstanding mathematicians on the continent was to be received in Cambridge only in the first decade of the nineteenth century.

The man upon whose writings this creative Cambridge tradition was founded was born at Woolsthorpe, near Grantham in Lincolnshire, on Christmas Day 1642, after the death of his father. The Newtons were lords of the manor, but their house was also a working farm. Isaac Newton showed no interest in running the family estate till late in life. Having been brought up by his mother during his first three impressionable years he was placed with his grandparents for the following seven and a half years, after her second marriage. This was not a happy time for the boy. After dame schools he was sent to the excellent Grammar School in Grantham, where (possibly) he first learnt mathematics as well as Latin. He also learnt somehow to use his hands. Reluctantly, finding the boy incurably bookish, his mother sent him to Trinity College, Cambridge, in the lowly status of a sizar and on a tight allowance. Although legend denies him early brilliance, Newton was

largely free to follow his own bent in study, and he was elected a scholar in his third year. At about the same time surviving notebooks portray him as engaged in the study of advanced mathematics, and reading recent publications on physical science, such as Robert Boyle's *Experimental History of Cold*. As Newton himself recalled in middle age, not inaccurately:

> In the beginning of the year 1665 I found the Method of approximating series and the rule for reducing any dignity [power] of any Binomial into such a series. The same year I found the Method of Tangents... & in November had the direct method of fluxions & the next year in January had the Theory of Colours & in May following I had entrance into the inverse method of fluxions [integration]... All this was in the two plague years 1665 & 1666 for in those days I was in the prime of my age for invention and minded Mathematics and [natural] Philosophy more than at any time since.

Later, in old age, Newton told a few friends the story of the apple falling in his orchard, which induced him to speculate about gravity. The last statement in the above quoted passage, at first sight implausible, may well be true. Within a year or two of his return to Cambridge in 1667 from his long visit to Woolsthorpe while the university was closed by the plague, after completing his early experiments on light, Newton became increasingly absorbed in the study of the chronology of ancient history, Christian theology, chemistry, and alchemy. Certainly the bulk of his surviving papers bearing on non-mathematical subjects is greater than those dealing with mathematical and exact scientific work, but they are not the subject of this contribution.

In all his intellectual activity Newton was essentially autodidact, and much of this lay outside ordinary academic paths. For example, his skill in experimentation was acquired by thought and practice: he was not the first in Cambridge to attempt experiments but he was by far the most exact and successful in the art. Further, he wove the results gained from precise experiments into both verbal and mathematical reasoning so tightly that he believed many of his conclusions

to be unassailable. Newton was not, however, singular in combining the highest mathematical abilities with an interest in theology; Isaac Barrow, first holder of the Lucasian Chair, wrote on theology as well as mathematics and was in Holy Orders, but Newton kept his unusual theological interests extremely private and took trouble to avoid the necessity for ordination. In the autumn, after his return to Cambridge, Newton was elected a Fellow of Trinity and two years later Barrow resigned his Chair, to which Newton was appointed. It is likely that Barrow meant to favour his younger colleague, not for the first time; certainly he introduced Newton to the London world of mathematics. Newton's reputation there was further enhanced when it became known that he had designed and built a novel telescope, a reflector. This success earned him election as a Fellow of the Royal Society at the end of 1671 and his first appearance in print. Not everyone was at first prepared to adopt the new ideas about light and colour, which had led Newton to avoid refractive lenses, and admittedly his first public explanation of them lacked clarity. In consequence, he was caught up for three years in a sharp debate embittering to his proud and sensitive nature. For his first professorial lectures Newton had continued where Barrow had left off, so that by early 1672 he had prepared a large and well-reasoned exposition of his ideas on optics and the nature of light, well meriting publication. Though Newton toyed with two or three schemes for putting the lectures into print, in the end he kept silent on his work for more than thirty years and the *Optical Lectures* were printed only after Newton's death.

After taking an early interest in the Royal Society, Newton in the mid 1670s withdrew from the London scene. His withdrawal may perhaps also be seen in his tendency, at this time, to neglect mathematics and natural philosophy while devoting more effort to his investigations of alchemy, theology, and early history. It would be rash, however, to assert that Newton ever completely abandoned any form of study that had once occupied his mind; and in every topic that he took up he was immensely thorough in mastering all relevant materials. It is ironic that a letter from his chief early critic, Robert Hooke,

asking about the shape of the true path in space traced by falling bodies, led him back into physical science, thus preparing the way for his greatest achievements. Their correspondence at the turn of the years 1679–80 impelled Newton to calculate that if an object were drawn or impelled obliquely towards a point with a force varying as the inverse square of its distance from the point, the line traced by the object must be an ellipse (or, at the Earth's surface, a parabola). Accordingly, when in August 1684 the astronomer and geometer Edmond Halley called upon Newton in Cambridge specifically to ask what curve would be traced by a planet moving round the Sun and always drawn towards it by such a force, he could at once reply that it would be an ellipse, as astronomers had already learned from Kepler. This pair of questions gave rise to the *Principles*. Soon after Halley's visit, in the autumn of 1684, Newton wrote – and transmitted to the Royal Society – an outline of the work on mechanics extending to nine pages. '[R]arely, if ever, in the whole history of science,' writes D.T. Whiteside, 'has more which is outstandingly original been expounded in fewer pages.' Not only did Newton in this short tract *On Motion* demonstrate the bare basics of a whole system of celestial mechanics, but he also outlined a mathematical treatment of the mechanics of fluids and of motions through fluids (such as air or water) which was to be developed in Book II of the *Principles*. The whole work (510 pages of Latin and geometry) is in three books, Book I setting out the abstract theory of motion, Book III the system of the world in which the general theory is applied to a variety of natural phenomena – the orbits of the planets and of their satellites, of comets, and of the Moon; the form of the Earth as shaped by its rotation; the motions of the seas; and it is shown that Newton's calculations of these phenomena agree with observations. All this vast amount of work was drafted and re-drafted in at most two years' writing time; Newton himself put it at eighteen months. Perhaps one is at first most impressed by Newton's unprecedented ability to apply his theory of motion and the principle of universal gravitation, to the data derived from the observations of astronomers, physicists, and travellers, to account with precise quantitative accuracy for a wide range of

phenomena of the grandest kind. But also of universal significance and timeless value were the fundamental concepts of space and time, force and motion, with the general principles to be adopted in natural philosophy, which are set out in the first pages of the first book and are both enriched and illustrated in the remainder of the work. Everyone has heard of Newton's Laws of Motion. The historian can follow the evolution of Newton's general principles from their appearance in the first draft of *On Motion* (1684) to their final forms in the last edition of the *Principles* (1726). Nothing can be more impressive than Newton's gradual approach in his manuscripts to a set of principles which were to dominate physical science for two hundred years and which are still valid for such purposes as the calculation of the orbits of man-made satellites.

The mathematics employed by Newton in the *Principles* was the geometry of his age with enrichments of his own. His language is always geometrical; when he employs infinitesimals they are lines and areas. Newton was always opposed to the solution of geometrical problems such as the movements of bodies by algebraic manipulations, saying that 'algebra was only for bunglers in geometry', even though he himself so greatly developed algebraic analysis by devising an algorithm for the use of infinitesimals. This he occasionally employed in the solution of particularly recalcitrant physical problems. It is this adherence to geometry, of course, which has made the approach of *Principles* difficult during the last century or more, unless it is 'translated' into a more modern idiom. It was once supposed (and Newton himself tried to promote this notion) that the author of the *Principles* had used analysis (the algebraic method of fluxions) to solve problems before working out the geometrical demonstrations that he published. Evidence to support this supposition exists in the case of only a very few propositions. Newton's geometrical conceptualisation was so powerful that he could work out the demonstration directly without preliminary analysis.

When the *Principles* was first published in 1687, Newton's researches in pure mathematics were entirely hidden from the public.

Indeed, that book itself contained the first hint of his method of fluxions. Even by 1700 this was known to a few close friends only: N. Fatio de Duillier, John Craige, and David Gregory. A pair of mathematical treatises by Newton at last appeared with *Opticks* in 1704; they were not for tyros. Meanwhile, from Hanover, Gottfried Wilhelm Leibniz printed in 1684 an essay outlining the differential calculus – equivalent to Newton's fluxions – of which he had formed the first notions about 1675, some six years later than Newton. The Leibnizian calculus was taken up and developed from about 1690 by French and German mathematicians particularly, for whom therefore Newton's treatise on the quadrature of curves contained nothing new. Meanwhile the personal relations between these two great men had been soured by claims that Leibniz's calculus had been modelled on Newton's older fluxions and the counter-claim that Newton had adapted his method of fluxions from an (unanswered) letter written to him by Leibniz on 11 June 1677 (long before anything of fluxions was in print), which indeed contains the basis and notation of the calculus. The dispute over the priority in invention between Newton and Leibniz, soon extended to other matters, continued even after Leibniz's death in 1716. It had not a little to do with the resistance to Newtonian natural philosophy in Germany, and caused a century's isolation of British mathematicians from their continental colleagues.

The *Principles* was at once warmly acclaimed in Britain, though Newton for some years had few active followers. Very few could master the book; its geometry was beyond John Locke who nevertheless adopted Newton's natural philosophy. Hence the significance of its popularisation effected from the 1690s onwards, by Samuel and John Clarke, David Gregory, William Whiston and John Keill, at first in Latin for the learned then in the vernacular. There were bizarre English critics of Newton, but long before Newton's death academic opinion had chosen his system and proclaimed Newton as a great national hero. The pomp of Newton's funeral, the splendour of his monument in Westminster Abbey, signified a height of public esteem that astonished foreigners, but in which they shared during the next generation.

Newton became the philosophical Moses who had finally dispelled medieval obscurantism and brought Europe into the clear day of the Enlightenment. Newtonian science became the model of all that was clear and certain, yet in perfect conformity with Christian principles. The thorns in Newton's thought had not yet pricked.

The publication of *Opticks, or a Treatise . . . of Light* in 1704 brought additional dimensions to Newtonian philosophy; this was a book begun (in Latin) as long ago as 1687, drawing on experiments first made in the 1660s. As Bernard Cohen has emphasised, *Opticks* was a book as firmly grounded by reason upon experiments as the *Principles* was grounded upon mathematics and observation (though in the earlier volume, experiment also has a place). *Opticks*, a relatively approachable volume, set out a body of Newton's natural philosophy that could be mastered without geometry; moreover, as the *Principles* furnished a model for mathematical philosophy (though in a mathematical formalism soon to be obsolete), *Opticks* provided a model that might be emulated in investigations quite different from that of light, such as those of electricity and chemistry. In this way Newton's non-mathematical writings – not forgetting such minor works as his study of heat (1701) and of the nature of acids (1710) – extended greatly the range of his influence upon contemporaries and posterity. There is, of course, this difference: the place of the *Principles* in science is, one might say, immortal and its interest for mathematicians eternal, whether in its original form or in that of an interpretation using calculus – an older one such as that by J.M.E. Wright, *A Commentary on Newton's Principia* (1833) or the recent work by Subrahmanyan Chandrasekhar, *Newton's Principia for the Common Reader* (1995). *Opticks* for scientists more obviously belongs to a past age. However, for philosophers and historians it is certainly hardly less enthralling than the *Principles*, for in *Opticks*, rather than elsewhere in print, Newton set out in the ever-enlarging group of *Queries* his deepest thoughts about the creation and nature of matter, the character of force or activity in nature, and the relation of God to his creation.

It is well known that Newton's fundamental notion of light

persuaded him that the correction of both spherical and chromatic aberrations by the perfection of lenses was (at least) very unlikely, and that he (for the first time) constructed a pair of model reflecting telescopes that promised well. The instrument which he presented to the Royal Society (December 1671) first aroused the interest of that body in Newton. Though attempts to make larger reflectors failed in Newton's life time, and later a way was found to correct the effects of lenses, since the middle of the nineteenth century the largest telescopes have always been reflectors.

The successful completion and almost rapturous reception of the *Principles* by his colleagues greatly altered Newton's life. His friendship with Edmond Halley and David Gregory grew warmer; besides sharing his work with them, he looked forward to further publications and began *Opticks*. Perhaps fortunately the progress of its initial Latin text was hampered by the Anglican response to the Catholicising policies of King James II, to whom nevertheless the *Principles* was dedicated. In the University's resistance Newton emerged as a leader. In the very month that Book III of the *Principles* was sent for printing to London, Newton appeared there before the infamous Judge Jeffries in the Court of High Commission as a delegate to answer for the University's failure to obey royal commands. Its refusal was one incident in the deposition of James and the accession of William III, warmly welcomed by Newton. When the latter offered himself as a candidate he was easily elected as a university member of the Convention Parliament of 1689–90. A year's living in London altered the pattern of Newton's life. He now longed to abandon Academe for some significant post in the capital. His wish was not quickly gratified, and it is possible that the mental strains associated with his disappointments caused a short episode of mental breakdown that he suffered in 1693, particularly associated with his new London friends. Nevertheless, in these years after his return to Cambridge, he embarked upon a new mathematical treatise (only printed in 1704), re-drafted *Opticks* in English, and began preparations for a new edition of the *Principles*.

At last in 1696 an appointment as Warden of the Royal Mint in

the Tower, with responsibility for a total recoinage of the English money, enabled him to leave Cambridge without regret. For the remaining thirty years of his life Newton lived the comfortable, indeed opulent, life of a London bourgeois; with his young, beautiful, and witty half-niece Catherine Barton as his hostess for much of this time, Newton entertained the aristocrats, politicians, and distinguished foreigners with whom his office and intellectual distinction brought him into contact, especially after he became Master of the Mint in 1700. Election as President of the Royal Society in 1703 – which he retained till he died – confirmed his powerful position.

In these London years, though accomplishing no new work but rather being much concerned with his own past, Newton was certainly not mentally inactive. He polished and published the long-delayed *Opticks*, adding to the book the speculative *Queries* already mentioned, much enlarged in later editions. The death in 1703 of Robert Hooke, his old critic who had infuriated Newton in 1686 by claiming priority in the notion of universal gravitation, enabled Newton to bring his book out in the following year. This had as warm a reception as the *Principles*, and one more widely spread since it could be read by non-mathematicians. The re-issue of the greater text, although it had been much revised in the 1690s and since, occupied Newton and his able co-editor Roger Cotes for the four years 1709–13. In his latter years Newton revised the text yet again, less profoundly, for a third edition in 1726.

A constant preoccupation of Newton's later years was his continuing quarrel with Leibniz and his younger allies. In 1699 Newton's friend Fatio de Duillier printed the accusation that Leibniz had based his calculus on Newton's prior (but unpublished) method of fluxions. Leibniz made a calm rejoinder, but when Newton himself printed two short mathematical treatises as appendices to *Opticks* (1704), one being *On the Quadrature of Curves*, Leibniz in a long review of the book insinuated that Newton (second into print) was the plagiarist. This infuriated Newton and introduced a long paper war, outlasting Leibniz himself into the 1720s. Though this was pursued by Newton with vigour and enthusiasm, occupying large amounts of his time, and

though most Englishmen remained for more than a century strongly his partisans, the calculus dispute added nothing of permanent value to Newton's writings nor did it increase his real international reputation, though (at first) it may have helped to make his name more widely known on the continent.

His fame abroad grew steadily during the last decades of his life, even in France which was so long divorced from Britain by wars. *Opticks* was accepted and admired comparatively rapidly. The English text re-appeared twice with Newton's revisions, while the Latin translation (1706) circulated on the continent where it was reprinted several times before 1750. There were in addition translations into French (Amsterdam 1720, Paris 1722). The early prejudice against Newton's theories of light and colour were overcome in his lifetime, especially after more than one attempt to repeat his essential experiments had succeeded. The *Principles*, or more exactly the book's central concept of universal gravitation according to the inverse square law, was long opposed by both the mathematicians and the natural philosophers of the continent who, besides their preference for the calculus of Leibniz, adhered to the Cartesian notion that all transfer of motion between bodies took place by contact. The *implied* corollary of universal gravitation that two bodies in empty space must exert forces on each other without any intervening agency (firmly rejected by Newton himself) was held to be so absurd as to render universal gravitation an impossible notion. In the five years immediately following Newton's death in 1727 some of the finest minds in Europe – notably Leibniz's adjutant, the mathematician Johann Bernoulli – strove to demonstrate that this basic principle was either ridiculous or redundant. As late as 1752, Newton's first biographer, B. le B. de Fontenelle, even better known for the most successful popular exposition of the Cartesian world view some fifty years before, published a book rejecting the Newtonian pragmatic use of gravitational attraction in favour of a revived Cartesian notion of space as being filled with an active aether causing the same effects.

Long before this, however, Johann Bernoulli and (in France) Pierre

Varignon had, during the first two decades of the century, 'translated' virtually the whole existing body of both rational and celestial mechanics into the language of the calculus, demonstrating many results first obtained by Newton. The first avowed, though cautious, follower of Newton in France was P.L.M. Maupertuis (1732); a few years later he proved by actual measurement that the Newtonian dynamic theory of the Earth's shape was valid. In this a colleague was A-C. Clairaut, another convert to Newtonian mechanics, who rendered Newton's incomplete theory of the moon vastly more accurate. With such works as these was raised the great body of research on mechanics undertaken (in France) during the second half of the century by J.L. Lagrange and P.S. Laplace, above all the latter's great *Mécanique Céleste* (1799–1825), while writings of parallel significance came from the younger members of the Bernoulli family and Leonhard Euler.

The eighteenth-century development of the legacy of the *Principles* was, in turn, the basis for much of the theoretical science of the following age, which, however, saw the abandonment not of Newton's experimental results in colour theory and optics but of his deepest ideas about the nature of light. The great observational discoveries of nineteenth-century astronomy seemed only to confirm the Newtonian view of the universe. Meanwhile, however, the immense growth of new knowledge in such experimental sciences as chemistry and electricity hinted at the possible future insufficiency of fundamental Newtonian concepts.

Further reading

I. Bernard Cohen. 1956. *Franklin and Newton*. Philadelphia, PA: American Philosophical Society.
I. Bernard Cohen. 1980. *The Newtonian Revolution*. Cambridge: Cambridge University Press.
Derek Gjertsen. 1986. *The Newton Handbook*. London: Routledge.
A. Rupert Hall. 1996. *Isaac Newton, Adventurer in Science*. Cambridge: Cambridge University Press.
Richard S. Westfall. 1980. *Never at Rest: A Biography of Isaac Newton*. Cambridge: Cambridge University Press.

Isaac Newton, born on Christmas Day 1642, was the son of a small landowner. Because of his fondness for books he was sent to Cambridge University (1661). At Trinity College, Newton attracted the attention of Isaac Barrow, Professor of Mathematics, who promoted his studies and in 1669 helped his succession to Barrow's own Chair.

Meanwhile, his Cambridge career had been interrupted by nationwide outbreaks of plague in 1665 and 1666, enforcing the dispersion of the University. Newton did not abandon the original investigations in mathematics and experimental optics begun at Cambridge; he continued them at his home (Woolsthorpe Manor, Lincolnshire). By analysing sunlight into a spectrum with a prism he discovered its composite nature, the cause of imperfection in telescopes, and by the autumn of 1668 had a theory accounting for the former and had made a reflecting telescope correcting the latter. A mathematical essay 'On Analysis' (?June 1669), sent to London in July, expounding his 'method of fluxions' and its geometrical applications brought forward Newton's name, but his mathematical discoveries only became open after 1700. By then Newton was already famous for his *Mathematical Principles of Natural Philosophy* (1687) expounding the mechanical theory of the universe. His gravitational theory too had begun at Woolsthorpe with an analogy between the fall of an apple and the constant 'fall' of the Moon towards the Earth.

Abandoning Cambridge in 1696 for the Royal Mint, Newton became Master in 1699. His second major work, *Opticks or a Treatise of Light* (1704) won applause at home and abroad. Lifelong Presidency of the Royal Society (1703) and knighthood (1705) made him Britain's most powerful scientist while his writings brought immortality. He died in March 1727 and was buried in Westminster Abbey.

A. Rupert Hall (b.1920), educated in Leicester and at Cambridge (where he founded the Whipple Museum of the History of Science) taught there and in the United States before settling as Professor of the History of Science and Technology at Imperial College, London (1963–1980). He is now Emeritus and a Fellow of the British Academy. His first published paper concerned a Newtonian manuscript, and (partly in collaboration with his wife, Marie Boas Hall), he has published a number of books on Isaac Newton. These include three volumes of *The Correspondence of Isaac Newton*, edited with Laura Tilling.

4 William Whewell: A Cambridge historian and philosopher of science

RICHARD YEO

The inclusion in this volume of William Whewell, the historian and philosopher of science, may require some comment. Unlike Isaac Newton, Charles Darwin, or James Clerk Maxwell, he was not a major scientific discoverer and does not feature in any list of great scientific minds. On the other hand, however, as a person who lived in Cambridge from 1812 until his death in 1866, Whewell's connection with that place was arguably more continuing and deeper than that of many others who began their scientific lives there. I shall begin with this last point and then return to Whewell's contribution to the historical and philosophical understanding of science, one that fully justifies his treatment in this book.

Born the eldest son of a Lancaster master carpenter in 1794, Whewell attended Heversham Grammar School from 1810 in order to compete for a scholarship to Trinity College. He was successful and was formally entered at Cambridge in 1811, beginning his first term in October 1812 as a sub-sizar. From this time, although he kept in contact with his family in Lancaster, he rarely returned there, preferring to stay in Cambridge with his books and his new friends. In spite of the plague in Cambridge in the spring of 1815, he told his sister that he had decided to stay, because the trip home was expensive, and 'because I can employ my time better here'. However, the young Whewell did not immediately appreciate the pleasures of the metropolis, describing his first visit to London in 1815 as a failure. He confessed to his sisters that he had only seen the city from 'the *outside*' because, not knowing anyone there, he could not 'see anything of its society'.

This assiduous avoidance of Lancaster and reluctance about

London did not stem from fear of travel. At a time when even local journeys were fraught with danger, Whewell was renowned as a bold adventurer: with Richard Sheepshanks, a fellow undergraduate, he survived shipwreck on his first attempt to visit the Continent in 1819; he travelled in 1823 with Kenelm Digby to study the architecture of churches and abbeys in Normandy and Picardy; and he visited Berlin, Freiburg, and Vienna in 1825 in search of the latest research on mineralogy and crystallography. Nor was Whewell ever content to restrict his travelling to obvious, and relatively, safe destinations. In the spring of 1826 he went to Cornwall with George Airy where they spent considerable time 1200 feet underground in a mineshaft experimenting on the mean density of the earth. Clearly relishing this performance (which he repeated in 1828), Whewell wrote to friends describing himself as a correspondent 'sitting in a small cavern deep in the recesses of the earth'. Above ground, and in the vicinity of Cambridge and beyond, his horse-riding style was remarked upon as reckless, if not cruel: Airy remembered a section of Cornwall as the place 'where Whewell overturned me in a gig'. Whewell climbed mountains in England, Wales, and the Continent well before this was deemed a virtuous, and gentlemanly, pastime.

But of all the places he visited (and avoided) Cambridge was the most important to Whewell. He regarded Trinity College as his true home, both personally and intellectually. This was where he made close friendships with John Herschel, the astronomer, Adam Sedgwick, the geologist, and Richard Jones, the political economist. Whewell was a devoted letter writer and his large correspondence comprises a significant network of scientific discussion with his friends and with men such as Charles Lyell, Michael Faraday, and James David Forbes, as well as with literary and philosophical authors. Moreover, he saw Trinity College as closely linked to the development of British and European science. In his daily life as an undergraduate, Tutor, and, from 1841, as Master of Trinity College, Whewell walked past the statue of Newton (by Roubillac) in the antechapel. As Master, he soon added another statue: a copy (by Henry Weekes) of the one of Francis Bacon at

FIGURE 4.1 Portrait of William Whewell by James Lonsdale (1825). *Source:* Trinity College, Cambridge.

St Albans. These two figures, both Trinity graduates, embodied for Whewell the vision and triumph of science, a heritage that had to be appreciated, taught, and defended. When the young Queen Victoria made the trip from London to Cambridge shortly after Whewell was appointed Master, this may well have been one of the conversation points as he escorted her around the college.

Whewell was a product of the Mathematical Tripos, established at Cambridge in the 1750s as a competitive examination held in the Senate House. The top group of students in this rigorous exam, usually extending over several days, were divided intro three classes, the highest being 'Wranglers', followed by senior and junior 'optime', while the remainder were called the 'hoi polloi' (majority) or simply 'poll men', and were deemed unworthy of honours. Graduating as Second Wrangler in January 1816, Whewell displayed the mental ability – and physical endurance – tested by the Tripos exam, and he was elected a Fellow in 1817 and appointed Mathematical Lecturer and Assistant Tutor in 1818 and Tutor in 1823. This training in Euclidean geometry and its applications in the mixed mathematical, or physico-mathematical, subjects of mechanics, hydraulics, optics, and astronomy were the basis of his scientific competency. It was also the centre of his vocation as a teacher and the occasion of his first appearance in print, as the author of *An Elementary Treatise on Mechanics* (1819). At this time he was also involved in discussions, led by Herschel, Charles Babbage, and George Peacock about the introduction of continental algebra and a new notation for the calculus. However, with pedagogic concerns foremost in mind, Whewell insisted that the Tripos concentrate on applied, rather than pure, mathematics, arguing that this would ensure that the utility of mathematics was demonstrated in the case of the most respected physical sciences. Fearing that students might learn mathematical formulae and processes without any reference to physical principles, he sought to retain the geometrical (or synthetic) proofs employed by Newton in the *Principia*, while acknowledging that the new analytical methods of Joseph Louis Lagrange's *Mecanique Analytique* were now the benchmark for specialists. In 1818 he told Herschel that if the reform of Cambridge mathematics went too far then, in a short time, students would be reading Newton's propositions merely as a matter of curiosity. His related concern was that students of science, and practicing men of science, should understand how major theories developed. By the 1830s, as his own scientific interests broadened beyond mixed mathematics to

mineralogy, crystallography, geology, and comparative anatomy, Whewell extended this historical perspective to all sciences. He began to do this in critical commentaries on the nature of science in his address to the British Association of the Advancement of Science in 1833, in two reports to that body on the state of mineralogy and mathematical theories of electricity, magnetism, and heat, and, for a wider audience, in some of the quarterly review journals of the day. In the latter he reviewed (albeit under the convention of anonymity) Herschel's *Preliminary Discourse on the Study of Natural Philosophy* (1830) and the first volume of Lyell's *Principles of Geology* (1830). He wrote these commentaries on recent scientific developments while he was assiduously investigating past science, a task that culminated in his monumental *History of the Inductive Sciences* (three volumes, 1837).

The *History* and its partner, *The Philosophy of the Inductive Sciences* (two volumes, 1840), constitute Whewell's major contribution to the understanding of science. I cannot give a comprehensive summary of these works here, but it is possible to ask how Bacon and Newton – Whewell's two Trinity College heroes – figured in his view of science. Newton caught Whewell's attention soon after his arrival in Cambridge. Writing excitedly to his father in 1814, Whewell explained that he had spent most of his allowance on a copy of the *Principia* – 'a book that I should unavoidably have to get sooner or later'. And it was achievements such as Newton's in the physico-mathematical sciences that became Whewell's standard of excellent science and the prime focus of his attempt to understand its historical development. Here Bacon became relevant because his rules of method – as given in his *Novum Organum* (1620) – had attained a certain orthodoxy in Britain. The so-called Baconian inductive method was regularly cited as the approved mode of scientific inquiry, and the reason for the success of men such as Galileo, Johannes Kepler, and Newton in the seventeenth century. In other words, these discoverers had taken up Bacon's call for a reform of natural philosophy, along empirical and experimental lines, and the world had reaped an intellectual harvest not witnessed by Bacon himself.

In his major works, Whewell's revised this relationship between Bacon and Newton. His first mention of Bacon in the *History* noted that this English Lord Chancellor had entirely missed the significance of Copernican astronomy. And, as he studied the history of science, Whewell came to the view that Bacon's account of induction was too mechanical and could never represent the actual thought processes by which great discoveries had been made. In using the term 'induction', Whewell, like Bacon, referred to the general process by which laws and theories were derived from observations and experiments; but he stressed that this was more than a mere generalisation from the facts and that, contrary to Bacon, there could be a productive role for hypotheses or guesses, even at an early stage of the inquiry. Whewell went even further and entertained a Baconian heresy; namely, the contention that the process of induction involved the addition of a conception from the mind of the scientist. He made this point explicitly in *The Mechanical Euclid*, published in the same year as the *History*: 'Some notion is *superinduced* upon the observed facts. In each inductive process, there is some general idea introduced, which is given, not by the phenomena, but by the mind'. Whewell suggested that, once this had been accomplished, previously detached observations assumed a unity that now required an imaginative effort to dissolve: 'The pearls once strung, they seem to form a chain by their nature'. In the *Philosophy* he put this formally, saying that when a number of facts were brought together under some conception, this was a 'Colligation of Facts'.

In Whewell's view of science, the mind was dynamic and creative; great discoverers were imaginative and speculative in their quest for knowledge of nature. Thus he regarded the speculative guesses of Kepler as an example of one path to such intellectual breakthroughs. There was no simple art or method of discovery, as Bacon's disciples liked to think, but Whewell contended that a philosophical understanding of science was possible on the basis of its history. Thus in writing about the events now known as the Scientific Revolution, Whewell spoke of the 'Inductive Epoch of Newton' as one which saw

major advances in physical astronomy and the theory of motion, culminating in Newton's spectacular feat of uniting terrestrial and celestial motion under the law of 'Universal Gravitation'. Whewell showed how Newton did this by building on the observations and theories of Galileo, Kepler, Tycho Brahe, and others, bringing these together to make 'the greatest discovery ever made'. So one implication from this history was that major discoveries are *prepared*. Whewell referred to a 'Prelude' in which facts and ideas were gradually clarified until they were brought together by the genius of an individual. His philosophical rendition of this was that certain 'Fundamental Ideas', such as time, number, cause, substance, likeness, and polarity, supplied 'Ideal Conceptions' appropriate to the various sciences: for example, that of the ellipse in Kepler's astronomy. However, it took an act of genius to bring apparently unrelated facts under such a conception, so much so that even Whewell felt unable to say much about this creative moment. Speaking of Newton he said that it was difficult to 'anatomise' the distinctive features of a mind capable of such inventive power. It was possible to identify key characteristics, such as mental tenacity, a clear grasp of the relevant ideas, and a tendency towards fertile generalisation; but beyond that Whewell considered it impossible to empathise with the mind of such a genius. Indeed, what he did say on this implied that in making his discoveries Newton walked close to the edge of an abyss. Repeating the reports about Newton's 'extreme absence of mind', Whewell offered this diagnosis:

> Often, lost in meditation, he knew not what he did, and his mind appeared to have quite forgotten its connexion with the body. . . . Even with his transcendent powers, to do what he did, was almost irreconcilable with the common conditions of human life; and required . . . the strongest character, as well as the highest endowments, which belong to man.

Whatever the moral costs of such a state, the imaginative leap made by Newton and other great discoverers could not be guaranteed by simple rules of method. Nevertheless, Whewell believed that the

conditions for such great discoveries could be understood in historical perspective. Here he renovated Bacon's contribution, suggesting that one such condition, was a vision of a new approach to scientific inquiry in contrast to that of Aristotelian scholasticism. Even before the fruits of such an approach were manifest, Bacon deserved the honour of being the first prophet of this new 'Experimental Philosophy', and the first legislator of the 'philosophical republic' in which it would be pursued.

Whewell wrote an intellectual history of science rather than one that stressed the influence of social factors. Nevertheless, he recognised the importance of institutions that supplied the material and social support for scientific inquiry. Thus from the mid 1600s, scientific societies, such as the Royal societies of London and Paris, created an environment in which individuals could experience a 'collision with other minds', thus ensuring that speculations were tested. Wealthy patrons had also been crucial to the support of sciences such as astronomy in which instruments were expensive; and enlightened princes had sponsored scientific expeditions. Here Whewell's attitude to Cambridge was complex. Although, like Oxford, this university was once a bastion of scholasticism, Whewell argued that it had been reasonably quick to embrace the new science, epitomised in different ways by Bacon and Newton. In the curriculum that Whewell inherited and sought to improve, mixed mathematics was taught in relation to examples from the work of Newton and other natural philosophers. This tradition secured the teaching of mathematics as the solid basis for scientific inquiry, giving Cambridge the reputation for advanced research of a distinctive kind, such as that represented in the nineteenth century by the physics of Kelvin and Maxwell. As mentioned above, in defending the place of mixed mathematics, Whewell was conscious of this legacy, but his attitude to the more general inclusion of science was seen by some as contradictory. He insisted that Euclidean geometry, Latin, and Greek formed the core of the undergraduate curriculum: both taught mental discipline and grounded the student in the heritage of Western culture.

Although he stressed the role of the Cambridge tradition,

Whewell's perspective was as a cosmopolitan, or at least, a European one. The breakthrough represented by the theory of universal gravitation, although it culminated in Newton, at Cambridge, was a European one, involving contributions from many countries. Other cultures, such as that of China and Arabia, in spite of boasting great minds and cultural wonders, had not produced a sustained scientific movement, partly because they had not fostered the kind of education required to instil the appropriate intellectual discipline, and also because they had not enjoyed free communication among individuals in different places. Yet, even within Europe, local traditions could be crucial. For example, Whewell was horrified by the way in which speculative philosophy – rather than geometry and the classics – had become so influential in the German universities during the careers of Fichte, Hegel, and Schelling. What appalled him was the way the teaching of these subjects, which lacked the definitive axioms and rules of classical studies, encouraged undergraduates to question the authority of teachers and to set up rival professors as embodiments of alternative views that could not be the object of definitive examinations. Here, then, an emphasis on location was crucial: Cambridge and its teaching practice were vital, in Whewell's opinion, to the continuity of the scientific tradition.

In affirming the traditional undergraduate curriculum, Whewell made a controversial distinction between permanent and progressive subjects. By the former, he meant geometry and the classical languages – subjects that rested on clear and indisputable concepts and principles. (As noted earlier, he was willing to add astronomy and mechanics for honours students.) In contrast, progressive subjects included the sciences, such as chemistry, botany, and geology (of which he was an ardent admirer), in which there was as yet no agreement on central theories. For this reason, although of great interest, these sciences were best left to be taken up later by those interested in pursuing science as a vocation, or as an avocation. In the view of his critics among men of science, Whewell's position was tantamount to the exclusion of the latest sciences from the centre of learning. Lyell protested that the

notion of waiting until the natural history sciences met Whewell's criterion for permanent status was an argument for teaching the present curriculum for another hundred years. A more positive way of looking at Whewell's conservative position is to suggest that he wanted Cambridge education to provide a general intellectual and moral foundation that cultivated citizens, not just scientists. That may still be a worthwhile aim.

Whewell's estimate of his own contribution to science is bound up with the lessons of his *History*. His self-assessment was frank, but possibly too harsh. In 1840 he told the geologist, Roderick Murchison, that 'there is nothing of such a stamp, in what I have attempted, as entitles me to be considered an eminent man of science'. Yet we need to note that Whewell was elected a Fellow of the Royal Society of London in 1820 and a Fellow of the Geological Society in 1827. He was appointed Professor of Mineralogy at Cambridge in 1828, in recognition of his mathematical papers on that subject and his acquaintance with the latest European research. And from 1833 he began to publish papers with John Lubbock (his former student) in the *Transactions* of the Royal Society on the tidal movements of the world's oceans. The aim was to produce a map of co-tidal lines showing the points throughout the globe where high water occurs at the same time. In 1837 he was awarded a Royal Prize Medal from the Royal Society for this research. His peers warmly acknowledged the work he did for the British Association in its early years, advising on the arrangement of the scientific sections and suggesting the idea of progress reports on various fields. Yet in his own estimate these achievements did not count towards high status in science, because unlike the achievements of Herschel or Faraday, for example, they did not represent major scientific discoveries. And it was these critical moments – their preparation, attainment, and development – that formed the dramatic structure of the *History*. In this work, the conception of science, as a sequence of heroic discoveries by great men, found one of its earliest and most eloquent statements. But, in affirming this perspective, Whewell relegated his own worthy contributions

to the sideline. Writing in 1869, in his *Hereditary Genius*, Francis Galton was able to judge Whewell's reputation in terms that have lasted until today:

> His influence on the progress of Science during the early years of his life was, I believe, considerable, but ... Biographers will seek in vain for important discoveries in Science, with which Dr Whewell's name may hereafter be identified.

Another lesson from Whewell's historical account was that science showed increasing specialisation over time. True, men such as Galileo, Newton, and Leibniz cast their minds across a wide intellectual domain, yet major discoveries required deep immersion in the appropriate facts and ideas of particular fields. After the work of the heroes of the seventeenth century, advances in science were made in a diverse range of subjects, some only recently recognised as specialist disciplines, such as electromagnetism. In fact, Whewell's suggestion of the word 'scientist' (by analogy with artist) at the 1833 meeting of the British Association was linked with his perception of ongoing specialisation. Following this up in 1834 in the *Quarterly Review* he announced the need for a term to describe all men of science, whether they be astronomers, chemists, botanists, or geologists, a name that would register a common identity in the face of possible intellectual fragmentation. Moreover, in spite of his own efforts to generalise about the history and philosophy of science, Whewell admitted that only experts in particular disciplines could fully grasp the relationships between the appropriate ideas and facts of their subject. This certainly applied to new subjects in which debates were still *sub judice*. It may be that Whewell's own self-confessed polymathic interests, which extended well beyond the physical sciences into moral philosophy, architecture, educational theory, natural theology, German poetry, and Greek translation, disqualified him from major participation in any one science. Indeed, Lyell told Whewell that he used to regret that he had not concentrated on one or two sciences, but now believed that by 'being a Universalist' rather than a specialist, Whewell was assisting

the progress of science through his proper 'calling' as an historian and philosopher of science. This view, shared by other members of the scientific community, was slightly different from the opinion of the Edinburgh wit, Sydney Smith, who famously remarked of Whewell that science was his forte and omniscience his foible.

Smith's epigram has affected Whewell's reputation, casting him as one of the last true polymaths, yet, for that reason, as one who made no memorable mark in any single subject. I prefer a more favourable judgement: that Whewell sought to balance the valuable traditions enshrined in Cambridge with an awareness that science was changing, and that communication, even among scientists, was becoming more difficult. Although he did not make a major discovery, Whewell practically founded the study of the history and philosophy of science. And in writing on these subjects he forged a vocabulary in which non-scientists could discuss science at a time when increasing specialisation of knowledge threatened to make such public conversations untenable.

Further reading

Yehuda Elkana (ed.). 1984. *William Whewell: Selected Writings on the History of Science*. Chicago, IL: Chicago University Press.

Menachem Fisch and Simon Schaffer (eds). 1991. *William Whewell. A Composite Portrait*. Oxford: Oxford University Press.

Isaac Todhunter. 1876. *William Whewell, DD. Master of Trinity College Cambridge: An Account of his Writings with Selections from his Literary and Scientific Correspondence*, 2 vols. London.

W. Whewell. 1837. *History of the Inductive Sciences, from the Earliest to the Present Time*, 3 vols. London. Also 3rd edn, 1857.

Richard Yeo. 1993. *Defining Science: William Whewell, Natural Knowledge and Public Debate in Early Victorian Britain*. Cambridge: Cambridge University Press.

William Whewell was born in Lancaster on 24 May 1794; he died in Cambridge on 6 March 1866. He was a Fellow and Tutor of Trinity College, Cambridge and held Chairs in mineralogy (1828–38) and moral philosophy (1838–55). He was appointed Master of Trinity in 1841. Known for his polymathic interests and range of writings, he is now most recognised for his *History of the Inductive Sciences* (3 vols, 1837) and *Philosophy of the Inductive Sciences* (2 vols, 1840).

Richard Yeo is Associate Professor (Reader) in the history and philosophy of science at the School of Humanities, Griffith University, Brisbane, Australia. He writes on the cultural aspects of science in the eighteenth and nineteenth centuries. Recent books are *Defining Science: William Whewell, Natural Knowledge and Public Debate in Early Victorian Britain* (1993), *Telling Lives in Science: Essays on Scientific Biography*, co-edited with Michael Shortland (1996), and *Encyclopaedic Visions: Scientific Dictionaries and Enlightenment Culture* (2001), all published by Cambridge University Press.

5 Adam Sedgwick: A confident mind in turmoil

DAVID OLDROYD

Adam Sedgwick, geologist, was born in the village of Dent in the Yorkshire Dales on 22 March 1785. He died in Cambridge in 1873 at the age of eighty-eight. Translated from the Latin, his epitaph at Norwich Cathedral reads: 'In Christ. To Adam Sedgwick, a Master among Philosophers, the Friend of Princes, the Delight of Little Ones, as One who Extended the Frontiers of Science, and was Fired with a Right Royal Love of Truth, whose Character was a Grand Simplicity, and whose Rock was the Faith of Christ, to Him, Once a Canon of the Church of Norwich, this Memorial is Raised by the Dean. 1873.' Sedgwick would surely have wished to be remembered thus. Does his epitaph offer a fair assessment of the man? What was his character? How did he behave and think?

Sedgwick's family in the Dales can be traced to the thirteenth century. His father Richard was Vicar at Dent, which was a reasonably prosperous place in Adam's youth, in the days of the hand weavers and spinners; but, never industrialising, it declined in the nineteenth century, and it now does well chiefly by virtue of the tourist trade, the great granite Sedgwick memorial stone in the village centre being a major attraction, said to epitomise his character.

Adam was third of a family of seven. He attended the small 'grammar school' in Dent until he was sixteen and then had two years at the more prestigious Sedbergh School nearby. From there, with the benefit of tuition from a notable local amateur mathematician, John Dawson of Garsdale, he obtained a scholarship (sizarship) to Trinity, where he studied mathematics particularly, being named Fifth Wrangler in 1808. Despite failing to gain the coveted first place, one of the examiners opined that Sedgwick had the greatest 'inherent power' of the candidates.

Following further examination, Sedgwick was elected to a College Fellowship in 1810, and taught undergraduate mathematics. He was ordained at Norwich in 1817, the last year in which he could keep his fellowship without taking holy orders. He was, to be sure, a deeply religious man – even if his choice was, in all probability, influenced by the attractions of a permanent place in delightful Cambridge, rather than a return to the relative intellectual solitude of a country parsonage and the teaching of schoolboys, such as had been his father's life. Along with his career decision, Sedgwick also committed himself to bachelorhood and celibacy.

Though he must undoubtedly have been a gifted mathematician, Sedgwick was not perhaps cut out for a career in that field. He found his role as a crammer of students burdensome, and his health broke down as a result of the efforts he made to secure his wanted fellowship. Though he recovered with the help of vigorous walking holidays in the Lake District and on the Continent, a pattern was set for most of his life: Sedgwick was amazingly fit and energetic when in the countryside, but became hypochondriacal and a valetudinarian when in Cambridge (or later Norwich) for extended periods.

Sedgwick is known (from fragmentary autobiographical notes) to have had some geological interests from an early age, and he pursued these interests on the Continent in 1816. Also, he was 'introduced' to the Geological Society of London in 1818. Even so, it is surprising that his scientific accomplishments were thought sufficient to secure the Cambridge chair in geology the same year. However, his membership of the largest college, where he was generally popular, helped him secure the position over a rival candidate from Queens'. He was elected FRS in 1821, John Herschel heading the list of those who nominated him. Sedgwick was President of the Geological Society in 1829–31, and President of the British Association when it met in Cambridge in 1833.

Starting from a low base of geological knowledge in 1818, Sedgwick immediately threw himself into the subject. He started his annual summer fieldwork in southwest and southern England, then worked his way northwards to Northumberland, and in the years

1822–4 he made the first systematic survey of the Lake District. He soon made the acquaintance of another young geologist, Roderick Murchison, and the two undertook a lengthy journey together round the north of Scotland in 1827, doing important reconnaissance work. In the 1830s, the friends collaborated in Devonshire and in Wales, where they began to unravel the structure and stratigraphy of the ancient Palaeozoic rocks of that Principality. Sedgwick 'invented' the Cambrian System, and Murchison the Silurian. In Devonshire they became involved in controversy with Henry De La Beche, first Director of the Geological Survey, from which acrimonious debate emerged the concept of the Devonian System.

The geologist companions Sedgwick and Murchison also travelled together on the Continent, but eventually they fell out about the question of the placement of the boundary between the Cambrian and the Silurian in Wales (each trying to extend 'his' System's domain at the expense of the other). This led to a titanic struggle for supremacy between the two geologists, both having some right on their side – Murchison correctly emphasising the importance of fossils for determining stratigraphic relations, while Sedgwick's forte was structural understanding. Both were indefatigable in the field and in the pursuit of their perceived scientific interests and reputation. In his lifetime Murchison's ideas largely prevailed, especially after he became Director General of the Survey, his nomenclature becoming embodied in the country's official geological maps. But later opinion has tended to lie on the side of Sedgwick's interpretations.

At Cambridge, Sedgwick gave an annual course of lectures, and built up the University's geological collections very considerably. His professorial salary was £100 per annum (£218 after 1821), but this had to cover the expenses of developing the Museum. His College fellowship also brought him several hundred pounds per annum (the amount varying substantially according to the health of the College's finances), but his geological fieldwork every summer had to be done at his own expense. So Sedgwick was not a wealthy man. It was partly for this reason, therefore, that he also – with the patronage of Lord Brougham –

took a 'second job' in 1834, as a canon at Norwich Cathedral. This might have been a sinecure, but it appears that Sedgwick took his responsibilities seriously, and resided, preached, and undertook ecclesiastical business in Norwich for several months each year, also encouraging the development of a museum in the city, where, as in Cambridge, he gave geological lectures. Indeed, Sedgwick was a strong supporter of amateur science and did much to assist the activities of the natural history society in Kendal, Westmorland, whose members reciprocated by fossil collecting for the distinguished Cambridge professor and making other useful observations.

Sedgwick was a renowned orator – or preacher and lecturer. Science lectures were not a required part of the Cambridge curriculum when he gave his first course in 1819, but by all accounts he attracted many students and dons to his lectures. His course was repeated until 1859 when he was seventy-four. He spoke extempore about general geological principles and the fieldwork that he had been doing recently, rather than pointless or tedious minutiae, and he used his museum specimens and maps and diagrams to explain his ideas. On one famous occasion, when the British Association gathered at Newcastle in 1838, he spoke to a group of savants at Tynemouth beach in the morning; and by the afternoon he had attracted a crowd of thousands, enthusing them with a heady brew of geology, political economy, natural theology, and patriotism, and reportedly drawing tears of emotion from many of his auditors. As John Herschel described it, Sedgwick 'led them on from the scene around them to the wonders of the coal-country below them, thence to the economy of a coal-field, then to their relations to the coal-owners and capitalists, then to the great principles of morality and happiness, and last to their relation to God and their own future prospects'.

Doubtless Sedgwick's Cambridge lectures contained similar elements, though this is not evident in either the printed syllabus or his surviving notes. Sedgwick also organised popular field excursions on horseback round Cambridge, everyone gallivanting round and concluding the day's proceedings at a suitable hostelry – an activity that

naturally increased 'student-numbers'! By common consent, Sedgwick was one of the great university teachers of his day. He is said to have had a grand eye, a grand face, and 'the beauty of a forest oak rather than that of a garden plant'.

By contrast, Sedgwick's predecessors at Cambridge had done little to enthuse anyone about geology (though before his time the subject had scarcely found its feet as a science that treated the *history* of the globe; and to be sure the Professor of Mineralogy, Edward Daniel Clarke, *had* got people excited about his high-temperature analyses of rocks and minerals). Sedgwick, in comparison with those who held the Woodwardian Chair before him, worked zealously for the development of Cambridge science, through his lectures, his stratigraphic research, and his enlargement and scientific description of the University's geological collections. He also played a central role in the foundation of the Cambridge Philosophical Society in 1819, the University's first significant scientific society, in whose *Proceedings* many significant papers have appeared.

Later in his career, Sedgwick was an important figure in curriculum reform. By 1845, he had risen to the position of Vice-Master of Trinity, under his old friend and sometime field companion the Reverend William Whewell, formerly Professor of Mineralogy and latterly master of everything. In 1847 the Chancellorship of the University fell vacant, and Sedgwick persuaded a group of Fellows, including Whewell, that Prince Albert was admirably suited to occupy the position, and could be expected to encourage science at Cambridge and reinvigorate the University. The Prince Consort was induced to have his name go forward on the understanding that he would be unopposed; but a group of St John's Tories put forward a rival candidate, the aged Lord Powis. Albert wished to withdraw, but he was persuaded to allow his hat to remain in the ring, with him being nominated 'without his consent'. An election followed, with Sedgwick strongly supporting the Prince's nomination; and in the outcome Albert was narrowly elected in the face of conservative prejudice and some anti-German sentiment.

Shortly thereafter Sedgwick was invited to the Royal residence in the Isle of Wight, evidently as a way of garnering support for Albert's plans for Cambridge reform, for the two had a lengthy discussion on university politics. The following year (1848), the Tripos examinations in Moral Sciences and Natural Sciences were established, and Sedgwick was soon setting examinations in geology. Pressure for change was coming from both within and without the University, Albert (and Sedgwick) favouring the former. But Parliament insisted on instituting a Royal Commission of Inquiry, and the new Chancellor went along with the idea. Sedgwick was invited to become a member of the Commission. He was reluctant to do so, given his previous preference for internal reform. But at Albert's urging he consented and in the event did much of the work, gathering evidence and largely writing the Report, which was published in 1852. The Chancellor had the geologist's ear, so that assuredly many of the Prince's ideas (such as the establishment of university science museums) found their way into the Commission's recommendations. These involved the establishment of Boards of Studies for each field, reform of examinations, fellowships and scholarships to be based on merit only, and in general a curbing of the powers and influence of the Heads of Colleges. The proposed Parliamentary Bill did not go far enough in this last regard, but on the urging of Sedgwick and three of his fellow Commissioners the Bill was emended and enacted in 1856, leading to the implementation of most of the proposed reforms.

We see, then, that Sedgwick was a man of liberal spirit (and, it should be said, of Whig politics) and passionate in his efforts to reform British education and develop science in British universities. Also, it should be remarked, he had risen from a position as third child of a remote country parson to become a major scientist and figure at Cambridge, received in the Royal Household, and privy to and instrumental in the implementation of the Prince Consort's designs for social and educational reform. At the same time, while moving in highest society, and intimate with poets such as Wordsworth and Southey, Sedgwick also maintained friendships with artisans,

especially those from the north with geological interests. Among these, the most important were the Keswick watch repairer and mountain guide, Jonathan Otley, who first assisted Sedgwick in understanding the mysteries of Lakeland geology in 1823, and the Kendal cobbler, John Ruthven, whom Sedgwick came to trust as a reliable assistant and fossil collector. Yet Sedgwick was also an ordained priest, and an inveterate preacher. How were these elements linked in his personality?

Apart from his technical writings, Sedgwick's major work was his *Discourse on the Studies of the University* (1833), the published version of a sermon preached at Trinity Chapel in 1832. This work, originally of eighty-two pages, grew like a snowball by the addition of various prefaces and appendices, until, by the fifth edition of 1850, it was a mammoth work of 766 pages. In a way this indicates something of Sedgwick's character. Like many teachers and preachers, and perhaps especially in the case of one who had risen so high and so successfully in the world, Sedgwick was always convinced that he was right, and he would defend his views with the utmost tenacity. There were four main issues on which he felt challenged: first, the question of the basis of morality; second, the question of whether there was any conflict between the findings of geology and essential Christian (Anglican) principles; third, the question of evolution; and, fourth, the status of 'his' geological system, the Cambrian. The first three of these challenges we see debated in the *Discourse* and its enlargements, which were stimulated by a trenchant review of the first edition by John Stuart Mill.

In Sedgwick's day, the influence of William Paley's natural theology reigned supreme at Cambridge. Students were expected to be thoroughly familiar with his *View of the Evidences of Christianity* (1794), and his *Natural Theology* (1802) was perhaps even more popular as *the* classic statement of the argument from design for the existence of God. The cogency or otherwise of Paley's design argument (or more sophisticated versions thereof) is something that continues to attract philosophical discussion to this day, but it is not a matter that need be reviewed much here. The aspect of Paley's

thought that chiefly exercised Sedgwick's attention and ire was his moral philosophy, which had a utilitarian basis. He held that 'utility is the touchstone of right and wrong'; which could be construed to mean that what was expedient was right. For Sedgwick, philosophical demonstration of God's existence was not really required: God was directly revealed to humans, through the Scriptures and through the example of Christ; and the knowledge of what was right and wrong was 'implanted' directly in the mind or in the conscience. So any sort of calculation of pleasures and pains, in the manner of Benthamite utilitarians, was an anathema. Sedgwick had no philosophical regard to the fact that as a teacher and preacher it was his role in life to be an 'implanter'. Moral relativism was unthinkable to him. Other 'implantations' – such as those of Muslims, for example – would have been simply wrong, and would need correction by appropriate preaching or proselytising. Sedgwick was, in fact, tantamount to a kind of Kantian absolutist in moral philosophy, though as a natural scientist he did not attempt to develop a formal philosophical argument in support of his position. It came from his religion, his reading and conversation, and his prayers, soaking in from the scholarly social milieu in which he moved.

Though Sedgwick's unsophisticated moral philosophy met with approbation in such places as Cambridge, Norwich, and Kendal, it was another matter with the London utilitarian, John Stuart Mill, whose review in the *London Review* (1835) made philosophical mincemeat of the distinguished Cambridge professor on numerous points. Sedgwick's real objection to utilitarianism was that it was supposedly a 'worldly' philosophy, rather than some spiritual source of morals. But there was a slide between two meanings of 'worldly': having to do with the non-spiritual; and having a penchant for things of the flesh, as I might put it. Thus Sedgwick the preacher engaged in philosophical sleight of hand.

On the question of the relationship between geology and theology, there *could* be no conflict, since both empirical science and Anglicanism were, so far as Sedgwick was concerned, obviously

correct – even though he must have known that some scientists disagreed with him, so not *all* scientific knowledge could be correct. I suggest that this supposed perfect interlocking of scientific and theological knowledge may explain the vigour with which he pursued his own knowledge claims against all comers. If he were shown to be wrong – if *his* observations were in error – then his whole metaphysical system would be in jeopardy.

So far as geology and religion were concerned, Sedgwick *assumed* (or thought he knew by revelation) that humans ('Man') were divinely created. So too were other species. And the results of stratigraphy seemed to show a series of new forms (species) successively appearing in the strata, and then dying out. Thereby species gradually came to resemble those we see today more closely. Whether new forms appeared simultaneously after some catastrophe, as most supposed in the early nineteenth century, or gradually and in piecemeal fashion as the influential geologist Charles Lyell supposed, was not a philosophically fundamental question. That was an empirical issue for the geologist to determine. Either way, each species was supposedly produced by a divine creative act of will, according to God's purpose. Hence the geologist could gain some insight into God's supposed creative acts, plan, and purpose. Geology was thus, for Sedgwick, a science that was not prejudicial to sound philosophy or religion. Rather, it had a kind of moral import.

In 1844, Sedgwick's picture of the universe was rudely upset by the anonymous publication (by Robert Chambers) of *Vestiges of the Natural History of Creation*, which suggested that although God had created the universe and its laws He had not had it under constant personal supervision subsequently, so that it had 'unfolded' naturally or evolved, in a kind of 'gestatory' process thereby reaching its present situation. This meant that 'Man' was not a special creation, and presumably did not, then, have a special kind of soul or moral conscience (though one might have thought this possible if the initial laws had been such as to yield in time 'emergent' mind, thought, and conscience). Perhaps it was the question of Man's immortal soul that was

the biggest issue. Cambridge, Norwich, Dent parsonage, etc., set much store by souls. If Chambers were correct, Man evolved from apes; and presumably human souls evolved from some simian precursorial mental/moral entity. Could *that* kind of soul go to heaven or hell? Could it have free will, and distinguish between right and wrong? Could it know God's will? It seemed exceedingly doubtful.

It was obvious to informed readers that there were errors in Chambers' science; but his book's table comparing the development (or appearance) of animal forms in the stratigraphic column with stages of embryonic development and the simple classification of animal forms had enough in it to attract close attention and to disturb a 'soulist' like Sedgwick. It was evident, then, that *Vestiges* was a book that could disturb undergraduates as well as Sedgwick, and might lead to that process which Victorians so dreaded: moral decay.

Faced with this challenge and evidently rattled, Sedgwick attacked, quite viciously, in his essay on *Vestiges* in the *Edinburgh Review* and again in the 5th edition of his *Discourse*. Privately, he wrote to Lyell:

> If the book be true, the labours of sober induction are in vain; religion is a lie; human law is a mass of folly, and a base injustice; morality is moonshine; our labours for the black people of Africa were works of madmen; and man and woman are only better beasts!

Here, I think, Sedgwick was all too correct. *Vestiges*, if true, did have such implications for Sedgwick's religion and metaphysics and for the social and ethical world in which he moved! His response was to attack the soundness of Chambers' science, which he was able to do to good effect.

Sedgwick was not happy either with the attempt of the popular Scottish writer on geology, Hugh Miller, to reconcile Genesis and geology in his *Testimony of the Rocks* (1857), by regarding each day of Genesis as having a corresponding geological era. Looking at the question as a geologist, Sedgwick knew that the details did not fit properly; and he rejected 'the scheme of stretching the Bible, like an elastic band,

till we can wrap up our hypotheses in its sacred leaves' (letter to Francis Close, Dean of Carlisle, 27 March, 1858, p. 4).[1]

Matters got worse for Sedgwick, though, when Darwin's *Origin of Species* was published in 1859. It had better science, and was even more naturalistic, materialistic (soul-less), and anti-teleological than Chambers' text; and the evolutionary link between humans and animals was more evident to the reader, albeit not explored in detail until the appearance of *The Descent of Man* (1871), by which time Sedgwick was perhaps past caring. Darwin was his friend, to whom he had taught the first rudiments of geology. He remonstrated in private: 'You have *deserted* – after a start in that tram-road of all solid physical truth – the true method of induction, and started off in machinery as wild . . . as Bishop Wilkins' locomotive that was to sail with us to the moon.' Parts, at least, of the *Origin* were, he opined, 'utterly false and grievously mischievous'. Sedgwick thought he could 'prove' that 'God acts for the good of His creatures'. He was wrong: his proof, had one been offered, would have involved a philosophical circle, from an assumption of a good God to the proof of a good God. The two naturalists ended up by agreeing to disagree. It was, of course, an insult to Darwin to suggest that he had abandoned inductive inference, though admittedly much of Darwin's empirical support for his theory was published after 1859.

How, then, did Sedgwick's geology work? In his early days, in the Lake District in the 1820s, he obtained topographic maps of the region, identified certain rock units, and systematically covered the region over three seasons, colouring in his maps according to the lithological units that he deemed to be important (actually given to him for the most part by his watch-maker friend Otley). Sedgwick did not look much for fossils, but he took many measurements of dips of strata and of cleavage (a distinction also made known to him by Otley) and of the alignments of beds, folds, faults, joints, and cleavage planes. The faults

---

1 Copy held at Dean Close School, Cheltenham. Photocopy kindly provided by Humphrey Osmond, School Archivist. My attention was drawn to this letter by Michael Roberts.

sometimes could be seen on the ground. At other times they became apparent when the different rock types were coloured on to the maps. His labour and energy were immense. Armed simply with hammer, acid bottle (to recognise limestones), map, compass, clinometer, and notebooks, Sedgwick tried to determine the structure of that immensely complicated region.

Trained in mathematics, the neophyte geologist was evidently attempting to see whether the strata displayed any recognisable regular geometric pattern. They hardly did, but when Sedgwick published his work he sought to subsume it under the general theory of the French geological theorist, Léonce Élie de Beaumont, according to which, as the earth cooled and contracted it supposedly formed a regular pattern of wrinkles in its crust, such that mountain ranges of similar age had similar alignments. The theory was fanciful and it never exerted much influence in Britain. Sedgwick soon gave up the idea; but the fact that as a young man he sought to deploy the French theory is significant, suggesting that he hoped to be able to formulate a geometrical (mathematical) theory of the earth. That he should have attempted to utilise Élie de Beaumont's theory is consistent with Sedgwick's training as a mathematician and with the Cambridge tradition that was being developed in the nineteenth century and which found expression in the *Proceedings of the Cambridge Philosophical Society*, namely that scientific knowledge should be quantitative and that science should be concerned with the search for laws, formulated mathematically.

But geology was not yet suited to such mathematical precision, and when Sedgwick tackled Wales in collaboration with Murchison in the 1830s, he adopted the more characteristic form of nineteenth-century geological reconnaissance research (which in Britain derived chiefly from the work of the surveyor/engineer/geological map-maker William Smith). That is, the procedure was to identify major lithological units in some selected area, identify their fossil contents, and then map boundaries, using fossils along with lithological criteria to extend

the observations from the 'type-areas'. As is well known, Sedgwick started in the mountainous area of North Wales, where he arrived at the idea of the existence there of a 'system' that came to be known as the Cambrian. It had some fossils, but they were scarce and not always very precise in their stratigraphic range. Murchison had an easier task. He started in Shropshire and the Welsh Border region, where the structures were less complicated and the strata more fossiliferous, the fossils being more precise in their stratigraphic ranges than those that Sedgwick was finding in the Welsh mountains. These Border rocks gave Murchison his Silurian System.

Eventually, the two geologists who started working as friends and collaborators fell out completely, for there was a debatable terrain betwixt the Cambrian and the Silurian, which both wanted to claim for 'his' own territory – the strata that years later were designated Ordovician. Sedgwick was adamant that his observations and structures were well founded, but during his lifetime Murchison's views largely prevailed. One of Sedgwick's papers was 'edited' by the Geological Society without his consent, and the bitterly offended Sedgwick broke off relations with the Society, and with Murchison, and withdrew to his redoubt at Cambridge, where the geologists working there upheld his views. (They were right to do so, for Murchison had made a serious mistake, conflating two rock units of different age – the 'Caradoc Sandstone' and the 'May Hill Sandstone' – which conflation was one of the main knots in the whole stratigraphic tangle.) It is scarcely surprising that Sedgwick reacted thus, having been forbidden by the Geological Society to read there any more papers on the classification and nomenclature of the older Palaeozoic rocks!

Sedgwick felt himself grievously wronged, and as regards his geology he became embittered in his old age. The great preacher and teacher could not bear to think that his ideas were rejected by anyone. His *cri de coeur* was published in the last year of his life in the unlikely place of the Preface to a *Catalogue* of the fossils in the collections in the Museum at Cambridge. There Sedgwick emphatically set out his

version of the history of the Cambrian/Silurian controversy, along with the elements of his creed. 'Man' was the 'last in the order of creation, and made in the image of the Author of his being' – not a product of the materialist process of evolution, a doctrine that Sedgwick regarded as a species of pantheism.

Sedgwick's egocentrism was written all over the Preface. Yet he was beloved by all those who knew him in Cambridge (if not in London), from the high and mighty to children and humble working men and women. He won a place in history – as founder of the Cambrian System; in the great Museum named and built at Cambridge in his honour after his death; and as a notable teacher and university reformer. All is rightly symbolised by the great block of granite in Dent that bears his name. His epitaph was fair, even if his philosophy and religion have not stood the test of time so well.

Further reading

J.W. Clark, and T.McK. Hughes. 1890. *The Life and Letters of the Reverend Adam Sedgwick, LL.D., D.C.L., F.R.S., Fellow of Trinity College, Cambridge, Prebendary of Norwich, Woodwardian Professor of Geology, 1818–1873*, 2 vols. Cambridge: Cambridge University Press.

J.M.I. Klaver. 1997. *Geology and Religious Sentiment: The Effect of Geological Discoveries on English Society and Literature between 1829 and 1859*. Leiden, New York, and Cologne: Koninklijke Brill.

[Mill, J.S.]. 1835. Professor Sedgwick's discourse: state of philosophy in England. *London Review*, **1**, 94–135.

D.R. Oldroyd. 1997. Adam Sedgwick and Lakeland Geology (1822–24). In *Comparative Planetology, Geological Education, History of Geology*, Wang Hongzhen, D.F. Branagan, Ouyang Ziyuan, and Wang Xunlian (eds). Utrecht and Tokyo: VSP, pp.197–204.

M.J.S. Rudwick. 1988. A year in the life of Adam Sedgwick and company, geologists. *Archives of Natural History*, **15**, 243–68.

J.A. Secord. 1986. *Controversy in Victorian Geology: The Cambrian–Silurian Dispute*. Princeton, NJ and Guildford: Princeton University Press.

A. Sedgwick. 1969. *A Discourse on the Studies of the University with an Introduction by Eric Ashby and Mary Anderson*. New York: Humanities Press and Leicester: Leicester University Press.

C. Smith. 1985. Geologists and mathematicians: the rise of physical geology. In *Wranglers and Physicists: Studies on Cambridge Physics in the Nineteenth Century*, ed. P.M. Harman, pp.49–83. Manchester: Manchester University Press.

C. Speakman. 1982. *Adam Sedgwick Geologist and Dalesman, 1785–1873: A Biography in Twelve Themes*. Broad Oak, London, and Cambridge: The Broad Oak Press Ltd, The Geological Society of London, and Trinity College, Cambridge.

Adam Sedgwick was born at Dent, Yorkshire (1785). After schooling in Dent and Sedbergh, he gained a scholarship to Trinity, Cambridge, and was Fifth Wrangler (1808). He obtained a College Fellowship (1810), took holy orders at Norwich (1817), and was elected Professor of Geology (1818). He also became a canon at Norwich Cathedral (1834), holding both positions until his death in 1873. From 1819, Sedgwick gave forty annual lecture courses, which stimulated much interest in geology at Cambridge. He undertook notable reconnaissance research, particularly in Palaeozoic rocks, and developed the University's geological museum and collections. He was elected Fellow of the Geological Society (1818), was Council Member for three periods, and President (1829–31). However, he severed relations with the Society in 1856 as a result of disagreements about publication of his papers on Palaeozoic stratigraphy. Sedgwick was a founder member of the Cambridge Philosophical Society (1819), and Vice-Master of Trinity (1845–62). He supported the work of the British Association, being President at its Cambridge meeting (1833), and thrice President of the Geological Section (1837, 1853, 1860). He was elected FRS (1821), gained the Copley Medal (1863), and honorary doctorates from Oxford (1860) and Cambridge (1866). He also did much work towards university reform, supporting the efforts of Prince Albert in this regard, assisting the Royal Commission that examined the affairs of the two old English universities, and writing much of its report (1852).

David Oldroyd, now retired, is Honorary Visiting Professor at The University of New South Wales. He took a Cambridge science degree in the 1950s, and then an M.Sc. in History and Philosophy of Science at University College, London. After secondary teaching in London and New Zealand he moved to a lectureship at UNSW in 1969, where he subsequently gained two doctorates and a personal chair. Specialising in history of geology, he has received awards from The Geological Society and The Geological Society of America and is Secretary-General of the International Commission on the History of Geological Sciences. A work in progress on the history of geology in the Lake District has led him to take a keen interest in Sedgwick.

6 Charles Babbage: Science and reform

ANTHONY HYMAN

Among Cambridge undergraduates there has long been an élite who have bypassed conventional studies to follow their own paths, usually to the dismay of their tutors, and gone on to make profound contributions to the sciences. Charles Babbage was an early member of this group. When Babbage went up to Trinity in 1810 Cambridge science was in a poor state. Experimental science barely existed, while mathematics languished, hampered by loyalty to Newton's inflexible dot notation for the calculus. The scene cried out for change, but as the transformation developed it became deeply involved with the contemporary political Reform movement.

Babbage and his friends fought for the systematic development of science and its application to commerce and industry on a national scale, but theirs was not the only ideology developing at the time. Against them was the curious chivalric revival which permeated the public schools and the Establishment. It has been observed that the ruling ideas of every age are the ideas of the ruling class, and Babbage and his friends lost the battle. The consequences of this defeat remain the subject of active discussion.

Babbage was born in London, south of the river in 1791. His family came from Totnes in South Devon, bankers who, like many other such families, had formerly been goldsmiths. A precocious mathematician Babbage was already well versed in the Continental mathematical notations when he went up to Cambridge. There he followed his own inclinations and ignored the standard curriculum. Instead he read deeply in the annals of the academies of Petersburg, Berlin, and Paris which were now available to him in the Cambridge libraries. He also purchased many volumes of French mathematics and science, even though they were expensive and difficult to obtain during the

Napoleonic wars. During his first two years at Cambridge Babbage was involved in a wide range of social activities, including whist, chess, sailing, and beagling.

His mathematical opportunity arose in a curious way. Babbage had had a partly evangelical education and in 1812 the evangelicals were involved in a controversy about whether the bible should be printed with explanatory notes or left as revealed perfection. The walls of Cambridge were plastered with posters and broadsheets circulated. Babbage drafted a parody urging that Lacroix's book on calculus was perfection and required no further explanation. The joke was taken up and the Analytical Society formed to discuss mathematics. The group of men who formed the Society, including John Herschel, George Peacock, and Edward ffrench Bromhead, formed Babbage's earliest mathematical associates. The Analytical Society published one work, *The Memoirs of the Analytical Society* (1813), which was printed by the University Printer. It was written by Babbage and Herschel, and the introduction placed great emphasis on the importance of mathematical notation, a theme which was to reappear continually in Babbage's later work. In a few years the Leibnitz notation had been adopted by Cambridge. The Cambridge Philosophical Society arose directly from the Analytical Society, and the Analyticals, as they called themselves, formed the core of the radical group which formed the Astronomical Club, forerunner to the Royal Astronomical Society, and led the National Science Reform group of the 1820s and early 1830s. During that period British science had a radical tinge which it has never quite regained.

Babbage was courting Georgiana Whitmore of Dudmaston in Shropshire, and through the Whitmores Babbage met Lucien Bonaparte who had taken refuge in England following a quarrel with Napoleon. Babbage learned much of the Continental approach to science and society from Lucien, and Babbage's views of science and engineering developed a breadth and incisiveness unrivalled in the England of his time.

Georgiana and Babbage were married from his father's house in

Teignmouth on the South Devon coast. They honeymooned in nearby Chudleigh and then stayed with Champernowne at Dartington Hall, a former royal estate near Totnes. They settled in Devonshire Street, St. Marylebone, in North London. The family grew apace while they spent their time peacefully on family visits or holidaying. Babbage also made a number of trips to the Continent, and in turn received Continental men of science, including, for example, Laplace, in Devonshire Street. Benefiting from his friendship with John Herschel's father and aunt, Sir William and Caroline Herschel, in 1815 Babbage lectured on astronomy for the Royal Institution, the privately funded research institute where Humphry Davy worked at the time. In this set of lectures he comes quite close to the modern idea that science advances by theory and refutation. Babbage became a fellow of the Royal Society. He continued to work in pure mathematics, developing themes which had been initiated in Cambridge. In the theory of functions he focused on improved notation, while in algebra he appears to have worked on a common pool of ideas developed by the Analyticals. He was also reading widely in both science and political economy. Then in 1822 Babbage put forward his plan to construct a Difference Engine for calculating mathematical tables according to the method of finite differences and printing the results of the calculations on-line, to use the modern term.

Babbage envisaged the calculation of tables of many classes, of both scientific and commercial interest, but he rested his public case primarily on tables to aid navigation. Tables were necessary for accurate coast surveys and positioning lighthouses as well as directly for use in navigation, while the tables then available were full of errors. If only a few ships were saved the cost of a table-making machine would soon be defrayed. For the world's leading seafaring country the case seemed obvious. The project secured the backing of the Royal Society and handsome government finance. However administrative machinery adequate to supporting such a complex and unprecedented technical project did not exist, and repeated difficulties were to arise with the government through inadequate documentation and similar problems.

In order to define his Engines, and also to assist in their development, Babbage developed his 'mechanical notation,' a unique combination of mechanical description and specification combined with logic diagrams. The notation went through many editions, gradually becoming a powerful tool on which development of the Engines entirely depended. It its later forms the notation comprised: a formal and systematic method for preparing and labelling complex, many-layered mechanical drawings; logic diagrams; and timing diagrams. Thus it provided not only a static description but a method for showing the entire *modus operandi* of the Engines.

A liberal in an era of Tory government when most public employment depended on political patronage, Babbage was unable to secure paid employment. However the Babbages had sufficient private means and lived contentedly in their house south of the recently enclosed Regent's Park. Babbage considered an invitation to become actuary to a life insurance company which was being formed. In the end he decided it would conflict with his Engine. The research he carried out was incorporated in a short book, *A Comparative View of the Various Institutions for the Assurance of Lives*, 1826, which was in effect the first consumer guide, though far better written than most of its successors. Then in 1827 two family events changed Babbage's life completely. His father died leaving him sufficient means for a very comfortable life and to support his costly research. A few months later Georgiana died leaving Babbage devastated.

Babbage's friends took him on a short trip to Ireland to visit work then in progress on the Ordnance Survey. His children were placed in the care of Georgiana's relatives while Babbage closed his house and departed on a journey through Europe which was to last eighteen months. In Frankfurt he met the son of the Tsar's coach-builder, and travelling with him to Munich Babbage learned enough of the subject to design his own highly successful caleche and have it built in Vienna a year later. He spent much time visiting men of science and renewed contacts with his Bonaparte friends. Babbage received early recognition on the Continent, was a member of the scientific academies, and

knew political leaders, including the Grand Duke of Tuscany, as well as liberal and intellectual figures.

After visiting Florence Babbage continued to Naples, exempt from customs thanks to his connection with the Grand Duke of Tuscany. Babbage undertook a brave personal exploration of the floor of the crater of Vesuvius, making measurements of temperature as he progressed. He also made a study of the Temple of Pozzuoli, concluding that the sea level had repeatedly risen and fallen. This was the sort of study that was contributing to the geological background for his friend Charles Lyell's classic *Principles of Geology*, whose first volume was to be published in July 1830. While he was in Italy, Babbage was elected to the Lucasian professorship: the emolument was insignificant, the duties light, and this apart Babbage remained outside the university system, but Cambridge had given Babbage its official blessing. On his long Continental tour Babbage became far more extensively acquainted with European science and manufacturing. Babbage's subsequent report of a meeting held in Berlin of the Deutsche Naturforscher Versammlung resulted in the foundation of the British Ass (The British Association for the Advancement of Science).

Babbage recrossed the English Channel in the autumn of 1828. He leased William Hyde Wollaston's former house in Dorset Street and threw himself into an extraordinary range of activities with a peculiar passion, as if trying to still the pain of his loss. This was the era of Reform. All the dreams of the liberals seemed capable of fulfilment and gathered about the demand for a parliamentary reform act. Babbage took an active part in parliamentary elections, including Cavendish's lively campaigns for the Cambridge University seat, which Babbage managed with Macaulay as his deputy. Later Babbage stood twice for election to the newly reformed parliament. He wrote an entertaining book, *Reflections on the Decline of Science in England, and Some of Its Causes* (1830), on the deplorable state of the Royal Society, a subject which had been close to the hearts of the Analyticals since their Cambridge days. He also wrote a highly influential book on political economy, a subject of great interest to all the liberal intellectuals of the

time, *On the Economy of Machinery and Manufactures* (1832), which was the first work fully focused on the factory instead of the farm. In the 1840s Babbage was to toy with what later became marginal value theory. I have long suspected that Babbage may have had a direct influence on William Stanley Jevons, who developed marginal value theory, possibly through meetings of the Statistical Society. However, I have never been able to prove the point. *The Economy of Machinery and Manufactures* was based on the conventional labour theory of value. Meanwhile Babbage's central interest remained his great Difference Engine. As design passed into manufacture he secured greatly increased financial support from the government, depending primarily on the backing of the then prime minister the Duke of Wellington.

After the Reform Act Babbage received no further support for his Engines, even though his friends among the Whigs were in power. The worldly and cynical Melbourne, prime minister after the Great Reform Act of 1832, was hardly the man to be interested in Difference Engines. However to have abandoned the project formally would have required effort, and it was so much easier to leave the Engine in limbo. Ultimately in 1842 the reforming Tory prime minister Robert Peel gave the project its quietus. Where in the 1820s support for a Difference Engine had seemed quite natural, now it seemed out of the question.

It was not solely a matter of the inclinations of individual prime ministers: the climate of opinion had changed. Many other schemes were abandoned during the same period, including universal education and any thought of systematic scientific and technical education. This was the high period of gothic revival, with the resurrection of chivalry, the *Return to Camelot* in full swing. The leading writer of the movement was Walter Scott, its leading politician the then young Benjamin Disraeli. The Eglington Tournament of 1838 was its high point. The interaction, or perhaps one should write 'conflict,' between these two movements was both complex and important and is fundamental to understanding the background to Babbage's scientific life. However, writers on chivalry show little interest in industrialisation, and *vice versa*, and the subject has not yet received the detailed study it

deserves. During the central period of his life Babbage was himself the leading figure of the scientific reform movement

In the nineteenth century it was common for historians to refer to 'the conflict between science and religion' during the scientific revolution. Modern historians of science have considered this description inaccurate, even though the actions of the Holy Roman Catholic and Apostolic Church have all too often seemed to justify it. If this nineteenth century view is seen to derive from the very real contemporary conflict between the section of the Reform movement associated with science, technology, and industry, on the one hand, and the reactionary chivalric revival with its close High Church associations, on the other, it becomes more comprehensible.

The chivalry revival movement was fundamentally opposed to everything Babbage stood for. For example Lord John Manners notoriously wrote:

>Let wealth and commerce, laws and learning die
>But leave us still our old Nobility!

This was Arnold of Rugby's heyday. Dr Arnold himself heartily disliked chivalric ideas, considering that chivalry exalted honour above justice and put personal allegiance before duty to God. He was also little interested in organised games. However, although the building of mock gothic castles and wearing of armour fell out of fashion, it was the adoption after Dr Arnold's time of chivalric ideas combined with muscular Christianity, and also with a deep-rooted suspicion of intellect and science, that, transmuted into the public school ethos, was to give the chivalric revival its lasting influence.

The difference between the attitudes to science in Britain and on the Continent is of fundamental importance and currently the subject of much debate. Even modest scientific education only came to England late in the century, and when science did develop in Cambridge later in the century, with legendary success, it did so in nearly complete isolation from industry. On the other hand the educational systems which developed on the Continent were inspired by

the Enlightenment. Ideas, and even institutions, had begun to develop before the French Revolution, but it was the post-revolutionary *grandes écoles* and the school system in France, as well as the spread of educational systems which followed Napoleon's armies, that established the pattern of education, such as the German technical schools. In Britain the educational system – or want of system – crystallised-out decades later, after the Reform Act. Moreover it was then a quite different historical period. In Britain industrialisation was well under way, and the class struggle was increasingly dominating discussion as the industrial proletariat matured. We must look not simply to philosophical considerations for the difference between the British and Continental education systems, but also to dislike of grey industry and fear of the developing proletariat which was now flexing its muscles.

In 1834, after a period of interruption in his work following difficulties with his workman Clement, Babbage returned to his Engine, and in the absence of government finance began to reconsider the project *ab initio*. These investigations led Babbage to launch on an extended period of the most intensive study lasting more than a decade. Gradually the concept of an Analytical Engine matured. The work continued until 1847 when he turned to the techniques he had developed for the Analytical Engines to develop simplified plans for a more powerful Difference Engine. It is this second Difference Engine which can be seen today showing its paces like a kinetic sculpture at the Science Museum in London.

Babbage's Engines included a remarkable range of ideas and concepts which were to re-emerge in the modern computer. A computer, or for that matter an Analytical Engine, is an abstract structure mapped on to a physical constructional system, but the abstract structure is not independent of the type of physical structure on to which it is to be mapped, and comparison between the Analytical Engines, which were to be built using Babbage's unique mechanical construction system, and electronic digital computers, is not straightforward. For example,

FIGURE 6.1 Printing and stereotype apparatus for Babbage's Difference Engine No. 2, constructed in 1991.
Source: The Science Museum, London.

when the main memory and the central registers have basically the same access and cycle times such concepts as scratchpad stores and cache stores have quite different meanings. None the less it is possible to identify in Babbage's Engines many techniques which were later to reappear in the modern computer, including: binary-coded store-addressing, a range of input and output systems, microprogramming, separate registers, and a program controllable by the results of previous calculations. Later, after Babbage had returned to the Analytical Engines in 1857, array processing was to be added. Just how close Babbage came to the concept of the general stored-program computer must in the absence of further information remain an open question. The details of Babbage's work were forgotten until well into the era of the modern computer, but through channels such as the nineteenth-century editions of the *Encyclopaedia Britannica* rather more was known about the Analytical Engines than is often realised. It is through such channels that one must look for Babbage's undoubted influence on Hollerith, and most probably also on Alan Turing. It is also possible that Turing might have learned of Babbage's work through the part III mathematics lectures and associated discussions: Cambridge has a way of remembering its own, and Babbage was never entirely forgotten in the University.

Between 1834 and 1847 Babbage was not considering actually constructing an Analytical Engine. After his difficulties with the first Difference Engine he was all too aware of the expense which would be incurred, and an Analytical Engine was not completed during Babbage's lifetime. However he carried out extensive feasibility studies, constructing trial pieces as he went along. Little remains of these experiments, but today there is little doubt that from a technical point of view construction of a working Analytical Engine was quite feasible. The successful construction of the second Difference Engine by Doron Swade at the Science Museum encourages the hope that it will one day be joined by a working Analytical Engine.

The Ninth Bridgewater Treatise, which Babbage published in 1836, was primarily concerned with presenting Babbage's idea of God

as Heavenly Programmer. Miracles were merely subroutines called down from the Angelic Store. This was mechanistic determinism pushed to its limit. In Babbage's writing we find also two basic concepts of the science of the era marching hand in hand: the commodity controlled atomistically by market forces, and the Newtonian particle moving freely in fields of physical force.

During the 1830s Babbage was much involved in the British Association for the Advancement of Science and was one of its original trustees. He was instrumental in starting the Statistical Section, and from this the London Statistical Society, later the Royal Statistical Society. Babbage also played a major part in the Battle of the Gauges, the argument about Isambard Kingdom Brunel's wide gauge. Babbage conducted a series of studies on a wide gauge train, and showed that at that time it was decisively superior to the Stephensons' narrow gauge trains. Babbage's intervention gave the wide gauge a further lease of life. His studies may fairly be claimed as the beginning of operational research. Babbage was instrumental in arranging for displays of the products of local manufacture during the 1838 and 1839 meetings of the British Association, and these exhibitions were direct precursors of the Great Exhibition of 1851.

After 1848, when work on the Engines had ceased and his draughtsman Jarvis had gone abroad, Babbage was freer to indulge in a wide range of scientific activities. He also went frequently into society. Babbage continued a lifelong interest in code-breaking, of which he one of the great theoreticians. Indeed so extensive was his activity that it seems to require further explanation than is directly available. Some have wondered whether Babbage was working regularly as codebreaker for the Foreign Office, or his friend Beaufort at the Admiralty. It might be interesting if those with access to the details of the Admiralty's code-breaking activities prior to the Second World War were to compare the techniques then in use with Babbage's methods.

The central event of the period was the Great Exhibition of 1851. Babbage wrote an entertaining book, *The Exposition of 1851, or Views of the Industry, the Science, and the Government of England*. His

description of the origin of the Great Exhibition is a classic. Babbage was himself proposed to organise the Industrial Commission for the Exhibition, but the government found him unacceptable, still no doubt at odds with him over the Calculating Engines. In the hands of mediocrity the Industrial Commission sank in importance, and in the end the Exhibition, with its overwhelming surfeit of ornament, was more in tune with the gothic revival than the world of science and industry.

During the Great Exhibition Babbage placed in an upstairs window a light with a shutter, forming an occulting light which flashed messages to passers-by. His friends would occasionally decipher the messages and return them to him. This little apparatus established the basis of the system of occulting lights which was later adopted for use in lighthouses all around the world. In 1854 Babbage's youngest son Henry Prevost returned from India on furlough and stayed with Babbage in Dorset Street, bringing youthful vitality to Babbage's widower's establishment. In 1856 Henry and his wife Min returned to India where Henry was to play a brave part in the Indian Mutiny. For Babbage life in Dorset Street was becoming lonely.

Finding himself alone again Babbage launched a new phase of work on the Analytical Engines. The work was in no way comparable to the systematic activity of the principal phase of work between 1834 and 1847. However there were some notable innovations. The store was laid out on an *X-Y* grid, and it is tempting to imagine Babbage conceiving the idea as he wandered through the stands at the Great Exhibition. He was toying with ideas for two, four, and multiple-processor systems, and then the grid layout led him to the concept of array processing. Babbage also devised a novel casting technique, seeking to bring the cost of construction within his own means. In 1865 Babbage published his felicitous memoirs *Passages from the Life of a Philosopher*.

Babbage died in 1871. Joseph Henry of the Smithsonian, who had visited Babbage in 1837, visited him again in 1870 and found his mental powers little diminished. After Babbage's death Joseph Henry observed that Babbage had contributed greatly to narrowing the gap between

science and practical engineering. However, successive governments had shown little interest in Babbage's plans for the systematic application of science to commerce and industry.

It is worth emphasising just how comprehensive Babbage's plans were: *inter alia*, universal education, extensive scientific and technical education at all levels, extended support for science and for technical research and development, operations research, scientific management, and the systematic application of statistical methods over the widest scope of activity. But Babbage and his allies' plans had been ignored. By the time Babbage died more people were beginning to understand the fundamental weaknesses in British industry, which were the achievement of the public schools and the scientifically illiterate Establishment they educated.

As the great pioneer of computing, operational research, and code-breaking, Babbage appears a very modern figure, but he was always a natural philosopher rather than the modern idea of a professional scientist; and it is the contradiction between Babbage's worldly and optimistic Enlightenment approach and the almost 20th century nature of his science that gives his life its special character.

Further reading

Ole Immanuel Franksen. 1993. Babbage and cryptography. Or, the mystery of Admiral Beaufort's cipher. *Mathematics and Computers in Simulation*, **35**, 327–367.

Mark Girouard. 1981. *The Return to Camelot, Chivalry and the English Gentleman*. New Haven, CT and London: Yale University Press.

Margaret Mary Gowing. 1977. Science, technology and education: England in 1870. The Wilkins Lecture, 1976. *Notes and Records of the Royal Society*, **32**, 71–90.

Anthony Hyman. 1982. *Charles Babbage, Pioneer of the Computer*. Princeton, NJ: Princeton University Press, Oxford: Oxford University Press.

Anthony Hyman. 1989. *Science and Reform, Selected Works of Charles Babbage*. Cambridge: Cambridge University Press.

Anthony Hyman. 1990. Whiggism in the history of science, and the study of the life and work of Charles Babbage. *Annals of the History of Computing*, **12** (1), 62–67.

L.S. Pressnel. 1956. *Country Banking in the Industrial Revolution*. Oxford: Clarendon Press.

Scientific Sketchbooks of Charles Babbage, in the Science Museum, London. For marginal value theory, see the Travelling Sketchbook. A crucial folder of Babbage's papers after 1857 appears to have been mislaid in the Science Museum's voluminous collection, making discussion of the question of Babbage's approach to the stored program difficult, and references cannot at present be given.

Charles Babbage FRS (1791–1871) was born in London South of the River in 1791. He was educated at small schools in: Alphington near Exeter; Winchmore Hill, North of London; Cambridge; and briefly by a tutor at Totnes Grammar. Between 1810 and 1814 he was at Cambridge, and in 1812 with his mathematical friends formed the Analytical Society which succeeded in introducing Leibnitz's mathematical notation to Cambridge. After coming down he married Georgiana Whitmore, and they settled in Devonshire St in North London. They had eight children, of whom a daughter died in her teens and three sons survived to maturity. In 1816 he was elected Fellow of the Royal Society. In 1820 Babbage was one of the founders of the Astronomical Society, later the Royal Astronomical Society.

In 1823 he started on a government-sponsored project to build a Difference Engine to calculate and automatically to print numerical tables, including logarithms and tables for navigation. In 1827 his father died leaving Babbage well off and able to finance his own research. Later in the year his wife died and Babbage left on a year and a half long trip to Europe. In 1828 he was elected to the Lucasian professorship at Cambridge, Newton's former Chair, which he held until 1835.

Late in 1828 he returned to England and settled in Marylebone. Here he ran celebrated soirées for thirty years. During the Reform period he campaigned for liberal candidates, and twice stood for election to the newly reformed Commons. Babbage campaigned actively for reform of the Royal Society and in 1832 published a major work on political economy.

In 1834 he started work on more general calculating engines, which led to the great series of plans for Analytical Engines. Babbage played a seminal part in launching the British Association for the Advancement of Science in 1831. Babbage also launched the Statistical Section of the Association, and developing from it the London Statistical Society, later the Royal Statistical Society. In 1838 he made a pioneering operations research study of Brunel's wide-gauge railway.

In 1840 and again in 1841 Babbage made presentations of the Analytical Engines to groups of natural philosophers in Italy. In 1847 he put aside work on the Analytical Engines and used the techniques he had developed while working on the Analytical Engines to prepare a new design for a Difference Engine.

During the following years Babbage worked on a wide range of subjects and invented the opthalmoscope and the system of occulting lights later adopted for use

in lighthouses. Babbage also worked extensively on cryptography, pioneering mathematical cryptography and apparently running a private Bletchley Park for his friend Beaufort at the Admiralty. In 1851 he published an entertaining account of the origin of the Great Exhibition.

In 1857 Babbage returned to work on the Analytical Engines, seeking to make manufacture sufficiently inexpensive to build an Analytical Engine himself. In 1864 Babbage was instrumental in securing the passage to a Bill to curb the activities of street musicians who had become a public nuisance. Also in 1864 Babbage published his felicitous biography, *Passages from the Life of a Philosopher*.

Charles Babbage died in 1871 and was buried in Kensal Green cemetery.

(R.) **Anthony Hyman, MA, Ph.D.**, was born in London in 1928 and was educated at Dartington Hall School and Trinity College, Cambridge. He worked for the ITT on early transistors, making a diffused transistor with his own hands in 1951. He carried out basic research on elemental linear polymers, and later worked on computer research for English Electric Leo Computers. He became a free-lance computer consultant and historian of science. Anthony Hyman held the Alastair Horne Modern History Fellowship at Oxford to carry out research for the biography of Charles Babbage and has written widely on the subject.

7 Charles Darwin
PETER J. BOWLER

Charles Darwin came up to Christ's College, Cambridge, late in 1827 to read for a BA degree. His intention was to prepare for ordination into the Church of England. He eventually obtained a rather undistinguished degree, but by then he had become determined to devote himself to natural history. Cambridge thus played a key role in turning Darwin to science, and it was his Cambridge contacts who arranged for his voyage on the survey vessel HMS *Beagle*, the event which changed Darwin's life completely. In the years following his return to England he developed the theory of evolution by natural selection, eventually published in the *Origin of Species* in 1859. The resulting debates made Darwin world famous. When Cambridge awarded him an honorary degree in 1877, the undergraduates in the audience dangled the figure of a monkey from the balcony. They, at least, appreciated the lesson of the *Descent of Man*, Darwin's analysis of human origins published six years earlier.

Historians all agree that the Cambridge years were a vital part of Darwin's development as a scientist. Although not actually studying science, he immersed himself in extra-curricular activity devoted to natural history and geology, and by the time he went aboard the *Beagle* he had a fair degree of competence in both areas. But the Anglican ethos of the Cambridge scientists was also a source of tension in Darwin's later life. Those biographers who adopt a sociological approach to the development of Darwin's ideas stress the contrast between conservative Cambridge and the radicalism of Edinburgh, where Darwin had previously made an abortive attempt to train in medicine. They also note how the development of Darwin's own theory put him into the radicals' camp, at least in principle, thereby undermining all the values of the Cambridge network which had giving him his place in the scien-

tific community. Cambridge was conservative in religion but Whig or liberal in its politics. Many of the dons favoured reforms that would give more power to the middle classes whose industrial enterprises were transforming the British economy. It was by applying Whig ideology (in the form of Thomas Malthus' population principle) to nature that Darwin created a theory that would undermine his belief in divine creation. Malthus' *Essay on the Principle of Population* argued that suffering and starvation were inevitable in a world where there would never be enough food to go around. Malthus himself saw this as an argument against the possibility of progress, but thought that the threat of suffering was God's way of forcing us to be hard-working and thrifty. Darwin took the element of the 'struggle for existence' from Malthus and turned it into the driving force of evolution, thereby eliminating the need for divine supervision of nature. The resulting conceptual tensions go a long way toward explaining the length of time it took Darwin to face up to the publication of his theory. They also explain the contrast between those who rejected the theory as a challenge to their religious beliefs and those who welcomed it for its application to the ideology of 'progress though struggle.'

At Edinburgh, Darwin had encountered the radical biologist Robert Grant, who used J.B. Lamarck's ideas about the transmutation of species to challenge establishment values. Lamarck's *Philosophie zoologique* of 1809 had argued for progressive evolution, but had also explained adaptation in terms of the inheritance of characters acquired by organisms in their efforts to cope with changed environments. For the radicals, this was the basis for a claim that future social progress was inevitable: people could improve themselves if given the chance. Darwin was no radical, and he did not accept the idea of evolution at this point, but he worked more closely with Grant than he would later admit in his autobiography. When he went up to Cambridge he seems to have been sincere in his belief that the world was divinely planned and created, although the seeds of doubt may already have been planted. Cambridge reinforced the more conservative side of his character – after all, his family was wealthy if not aristocratic, and had

much to lose if the radicals and revolutionaries succeeded in overturning traditional social conventions. He studied classics, divinity, and mathematics, but not the natural sciences (indeed one could not get a degree in science at this time). Darwin later claimed to have wasted much of his time at Cambridge on shooting and having a good time with his sporting companions. But in fact his outdoor activities had a more serious purpose: he became an avid collector of insects, inspired by his second cousin W. Darwin Fox. Fox also introduced him to the man who was to turn him toward a life in science, the Professor of Botany John Stevens Henslow.

It was Henslow who turned Darwin from a keen amateur naturalist into someone determined to devote himself to science. Darwin began to attend Henslow's lectures on a variety of botanical topics, which were supplemented by practical instruction and field trips. Later he was invited to the weekly 'open house' where scientific topics were discussed by eminent figures including Adam Sedgwick and William Whewell. Henslow evidently did his best to encourage the young man's natural inclination toward the study of nature and provided him with a sound, if extracurricular, scientific education. Darwin eventually became recognised as Henslow's leading disciple, and became a regular dinner guest with his family. Some of his fellow students felt that he was a little too obsequious, and Henslow too ready to acknowledge him as a favourite, but Darwin was responding for the first time to a mentor who would encourage him to participate fully in the intellectual and social world of science. The context of this training, though, was firmly within the tradition of natural theology. Henslow was a clergyman who saw no conflict between his religion and his science because he believed that species were divinely created. As part of his official studies, Darwin was required to read William Paley's *Evidences of Christianity*, but he also read Paley's *Natural Theology*, that classic statement of the argument from design. He was charmed by its long list of adaptations, showing how each species was designed by a wise and benvolent Creator. But under Henslow's tutelage Darwin became a 'philosophical naturalist' – he wanted to know

more about the laws which governed the complex operations of the natural world, although he still assumed that those laws were instituted by God. For all this theological conservatism, Henslow was a prominent Whig in his politics and supported the campaign for political reform. Here his views would also chime with those of the young Darwin, who came from a self-made family linked to the new industrial elite through the Wedgwoods, and who thus naturally looked to individual effort and initiative as the driving force of progress.

Darwin's more serious interest in science is demonstrated by his reading of J.F.W. Herschel's *Preliminary Discourse on the Study of Natural Philosophy*, an influential statement on the problem of defining the scientific method. He was also becoming fascinated by the tropics, copying out passages from Alexander von Humboldt's *Personal Narrative* of his explorations and observations in South America. By the time he took his degree (he was tenth in the list of those who did not go in for honours) in 1831, he was determined to travel south. For a while he entertained a plan to visit the Canary islands, possibly with Henslow. In the meantime, he had to spend two more terms at Cambridge, and it was at this point that he extended his interests to geology – a subject he had become disillusioned with at Edinburgh, thanks to the uninspired teaching of Robert Jameson. The Cambridge Professor of Geology was Adam Sedgwick, a rising star within the camp of 'catastrophists' who interpreted the transformations of the earth's surface in terms of violent upheavals and tidal waves. Sedgwick – also a committed Anglican – for a while accepted the popular view that the last great catastrophe could be identified with Noah's flood. But he was also one of the most enterprising workers in the application of the new principles of stratigraphy to the unravelling of the historical sequence of geological formations. He was just beginning the study of the ancient rocks of North Wales that would make his reputation through the establishment of the Cambrian system. Darwin travelled to Wales with Segwick and soon became an expert himself in the new techniques. It is significant that much of his time on the *Beagle* voyage was spent on geological observations, and

Darwin's early reputation as a scientist was built on the publication of his geological work.

In August 1831 an opportunity at last presented itself for Darwin to undertake the extended voyage on which he had set his heart. Captain Robert Fitzroy of the survey vessel HMS *Beagle*, which was returning for a second mission to chart the coastline of South America, let it be known that he was seeking a gentleman companion to accompany him as a naturalist to observe the geology and natural history of the region. The scientific network soon transmitted the request to Cambridge and to Henslow, who promptly recommended Darwin. After some objections from his father were overcome, Darwin let his name go forward and was accepted by Fitzroy. Bad weather delayed the departure, but on 27 December 1827 Darwin set off on the voyage that would change his life.

The five years that Darwin was away – much of it spent inland in South America while the *Beagle* tacked up and down the coast – yielded a mine of information and insights that would transform his worldview. He collected mineral specimens and especially fossils, sending them and his written comments back to Henslow, who arranged for publication. He studied the geological structure of the Andes and observed at first hand the effect of a major earthquake at Concepcion in Chile. Henslow had given him the first volume of Charles Lyell's *Principles of Geology*, warning him not to accept its 'uniformitarian' method which interpreted all change as slow and natural, requiring vast amounts of time. But Darwin's observations convinced him that Lyell was right: the earthquakes that plague the region are still elevating the Andes, step by step. He used Lyell's principles to develop a new theory to explain coral reefs in terms of the equally gradual sinking of the ocean bed. From this point on, Darwin was committed to explaining the present features of the earth and its inhabitants in terms of natural causes acting slowly over vast periods of time.

There were many discoveries related to the animals and plants of South America too. Darwin noticed that the fossil animals appeared to be close relatives of the present inhabitants, although the extinct forms

from earlier periods were often much larger. He discovered a new species of the flightless bird, the rhea, on the pampas of Argentina, and was led to wonder what geographical and environmental factors governed its distribution. How could two related species each be perfectly adapted to their own environment when they occupied overlapping territories? Experiences with the natives of Tierra del Fuego – three of whom had been brought back to England by Fitzroy and were now being returned – taught him about the life of the most primitive 'savages' and showed him that they could nevertheless be educated to observe European values and customs.

It was the Galapagos islands, several hundred miles out in the Pacific from the coast of Ecuador, which provided Darwin with most food for thought. Historians have now exploded the myth in which Darwin was supposed to have undergone a 'eureka' experience converting him to evolution when confronted with the different finches of the Galapagos. But he did notice that the various species of mockingbird inhabited different islands, and would later come to realise that the finches seemed to represent different adaptations to the various ways of getting a living on the islands. Shortly after returning to England, he was told that the various finches were distinct species, not mere local varieties of a single form. He began to feel that the conventional explanation of the Deity supernaturally creating species adapted to each environment broke down in these circumstances. Did it really make sense for Him to create a whole family of species adapted to such small and remote islands? Darwin soon became convinced that the species had evolved from an ancestral (probably South American) form that had established small founder populations on the various islands, each population then changing in a particular way to adapt to its new circumstances. From this point on he was a confirmed evolutionist.

The *Beagle* returned to England in 1836, and Darwin went back to Cambridge to unpack his specimens and consider what to do about publishing his observations from the voyage. He soon moved to London and for several years participated in the busy scientific world of the capital, becoming secretary and vice-president of the Geological

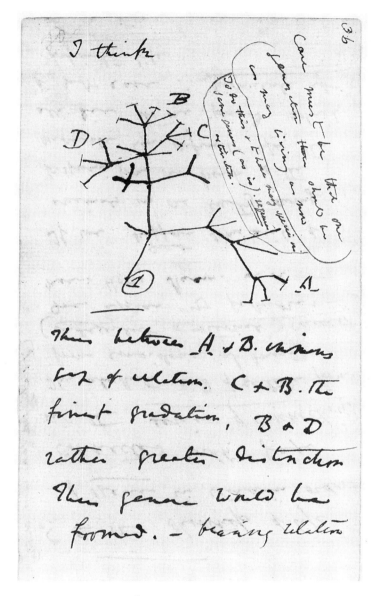

FIGURE 7.1 Diagram to illustrate branching evolution; from Darwin's Notebook B, p. 36 (Cambridge University Library).
Source: P.J. Bowler, Charles Darwin, the Man and His Influence, Cambridge University Press.

Society. His geological observations and theories were published, and various experts engaged to write descriptions of the natural history specimens. He married his cousin, Emma Wedgwood, and in 1842 the couple moved to Down House in the Kent countryside – Darwin was already beginning to experience symptoms of the illness that would make him a partial recluse for the rest of his life. He was not completely isolated, however; he could still visit London when he wished, and he built up an immense network of correspondents who could provide him with information relevant to his growing interest in transmutation. His ideas on this were kept secret at first, but eventually a number of naturalists were let in on the idea and invited to comment, the most active being the botanist Joseph Dalton Hooker.

The years in London were extremely active ones, and by 1842 Darwin was already able to write out a sketch of his new theory. His notebooks in which he recorded the observations and speculations that led him to the theory are preserved in Cambridge University Library and have now been published (there is also a major project to publish all of his massive correspondence). Having become convinced that evolution did occur, Darwin wanted to find a mechanism that would explain how the changes were produced. His investigations included extensive work with animal breeders, who were known to be able to produce massive changes in domesticated species. Darwin realised that there is considerable variation among the individuals making up any population and (to oversimplify somewhat) he discovered that the breeders effect changes by selection – they allow only those variants with the characteristics they need to breed, on the assumption that the individual characteristics will be transmitted to the offspring by heredity. Over many generations of continued selection within an isolated population, the desired character is fixed and enhanced. Darwin looked for a natural equivalent of this artificial selection, and eventually realised that it could be provided by the 'struggle for existence' caused by population pressure. Natural selection works because only those individuals with characters adapted to the environment will survive and breed. The key insight here was Malthus' principle of population, a classic product of

liberal, laissez-faire social thought. Malthus saw human misery as the consequence of overpopulation in a competetive society, and Darwin realised that the same force must operate among animals, resulting in starvation for those least able to cope with the environment. Historians have argued endlessly over the role played by ideological factors in Darwin's thinking. Some have tried to minimise it so that natural selection can be portrayed as a product of 'pure' science, but the majority now accept that the individualist ideology within which Darwin was steeped must have helped him to visualise nature along similar lines.

The new theory may seem like an extension of free-enterprise liberalism, but its consequences for the worldview of the respectable classes in Victorian Britain were immense. Not only did it demolish the idea of divine creation, making adaptation a natural process rather than a fixed state designed by God, but it also portrayed the process as driven by death and suffering – hardly the kind of system one would expect a benevolent God to set up. Nor was there any obvious goal toward whch evolution would be steered, since natural selection worked only by adapting species to their local environment. When samples from a single original population become isolated in a range of diverse habitats, as in the Galapagos finches, each sub-population will evolve in its own way and eventually their will be a group of distinct but related species. Darwin salvaged something by supposing that in the long run, evolution would favour higher levels of intelligence and adaptability, but from this point on he found it increasingly difficult to defend the view that the theory was compatible with belief in a benevolent Creator. Small wonder that he kept his ideas secret, although he was also trying to work out the detailed scientific implications of so broad a reinterpretation of the traditional worldview.

By the mid 1850s Darwin felt confident enough to begin writing a 'big book' describing his theory. This project was interrupted in 1858 by the arrival of a paper written by Alfred Russel Wallace, a naturalist then working in the Malay archipelago (modern Indonesia) outlining a theory similar to Darwin's own. Historians disagree over the degree of similarity – the traditional view was that Wallace had independently

discovered the whole theory of natural selection, but some historians (the present author included) think that Wallace's 1858 paper describes only a form of selection acting between varieties or subspecies, not between individuals within the same breeding population. Whatever the correct interpretation, Darwin himself felt that he had been forestalled and asked his friends Lyell and Hooker what to do. The result was the reading of the joint Darwin–Wallace papers to the Linnean Society, while Darwin rushed to complete the short, single-volume account of his theory which became the *Origin of Species*.

It was the publication of the *Origin* late in 1859 which sparked the great debate ending, a decade or so later, with most of the scientific community, and most of the educated public, converted to some form of evolutionism. Younger, radical scientists such as Thomas Henry Huxley welcomed Darwinism (as it soon became known) as a weapon in the fight to free science, and society generally, from clerical control. Conservatives, especially religious thinkers, found the theory difficult to swallow. It has to be confessed that the most spectacular confrontation, between Huxley and Bishop Samuel Wilberforce, took place not at Cambridge but at Oxford, at the 1860 meeting of the British Association. (Here too historians have exploded a myth – there is little evidence from contemporary sources of the sweeping victory conventionally attributed to Huxley.) Over the next decade the majority of scientists were converted, and even many religious thinkers. Many shied away from the most radical implications of the selection theory, however, preferring to see evolution as a purposeful, progressive process aimed inevitably at the production of humankind. Much of the evolutionary biology of the late nineteenth century owed little to the theory of natural selection, many 'Darwinists' looking for rival mechanisms such as Lamarckism (the inheritance of acquired characters) or saltations (evolution by sudden jumps). Even Huxley was lukewarm about selection and preferred saltationism. Darwin revolutionised his own world by converting everyone to evolution, but it was left for the twentieth century to develop the genetical theory of natural selection which has become dominant in modern biology. Meanwhile, various

forms of 'social Darwinism' became popular, based on a general enthusiasm for the idea of progress through struggle and often owing little to the details of Darwin's theory. The leading social Darwinist, Herbert Spencer, was a Lamarckian who subordinated the 'survival of the fittest' (his own phrase, not Darwin's) to the individual's ability to transform itself in the struggle for existence. Darwin himself eventually contributed to the debate over human origins with his *Descent of Man* of 1871. Here he anticipated the modern view that the first humans arose from an ape-like stock in Africa, probably standing upright before they began to develop enlarged brains.

Cambridge saw no great debate over Darwinism, perhaps because the Professor of Zoology Alfred Newton was an ornithologist, who was by no means opposed to the idea but saw no immediate way of applying it to his own research. Sedgwick, still active despite his age, was bitterly opposed. In the 1870s Cambridge became an active centre of research into evolutionary morphology, thanks to Huxley's protégé Francis Balfour (unfortunately killed in a climbing accident in the year of Darwin's death, 1882). Darwin sent his own sons to Cambridge, and thus came to visit fairly regularly in his later life. In 1869 he came for the graduation of his son, Frank, and explored the town, so full of memories – he felt that Cambridge was not the same without Henslow. The family met Sedgwick, who greeted Darwin as though he had not written the despised *Origin* and walked them round the Woodwardian Museum of geology until Darwin was prostrated (Sedgwick himself was eighty-six at the time). In November 1877 came the honorary LLD, with the whole family present in a Senate House full of rowdy undergraduates who cheered Darwin himself, and also the monkey dangled from ropes suspended from the gallery. Darwin met a lot of old friends and saw his old rooms at Christ's. He was still working actively on the phsyiology of plants, but was now slowing down. He enjoyed reasonably good health for a couple more years and published his last book, on earthworms, in 1881. In December of that year he suffered a heart attack which was the harbinger of a steady decline until his death on 18 April 1882.

Further reading

Peter J. Bowler. 1990. *Charles Darwin: The Man and His Influence*. Oxford: Basil Blackwell [now published by Cambridge University Press].

Janet Browne. 1995. *Charles Darwin: Voyaging*. London: Jonathan Cape.

Charles Darwin. 1845. *Journal of Researches into the Geology and Natural History of the Countries Visited During the Voyage of H.M.S. Beagle*. London. Reprinted London: Routledge, 1891.

Charles Darwin. 1859. *On the Origin of Species by Means of Natural Selection*. London. Reprinted Cambridge, MA: Harvard University Press, 1964.

Charles Darwin. 1871. *The Descent of Man and Selection in Relation to Sex*. London: John Murray.

Charles Darwin. 1958. *The Autobiography of Charles Darwin*, ed. Nora Barlow. New York: Harcourt Brace.

Charles Darwin. 1987. *Charles Darwin's Notebooks, 1836–1844*, ed. P.H. Barrett, P.J. Gautrey, Sandra Herbert, David Kohn, and Sydney Smith. London: British Museum (Natural History) and Cambridge: Cambridge University Press.

Adrian Desmond and James R. Moore. 1991. *Darwin*. London: Michael Joseph.

Charles (Robert) Darwin (1809–1882) was born in Shrewsbury, son of the medical doctor Robert Waring Darwin and grandson of Erasmus Darwin. The family had close links with the Wedgwoods, of pottery fame, and in 1839 Charles married his cousin, Emma Wedgwood. He was sent to Edinburgh to train in medicine, but abandoned this and went to Cambridge to obtain a BA in preparation for becoming a clergyman. Instead he became evermore strongly attached to natural history, thanks largely to the Professor of Botany, John Stevens Henslow. After graduating he set off as a naturalist on the voyage of the survey vessel HMS *Beagle* (1831–6), during which he spent much time in South America and also saw the biogeographical diversity of the Galapagos islands. On his return he was active in the Geological Society of London for several years before marrying and retiring to the country at Down House in Kent. Here he studied animal breeding and many other topics connected with the theory of evolution which he had just conceived. In the 1850s he began writing a large book on this topic, but after the arrival of A.R. Wallace's paper in 1858 produced the shorted version published as the *Origin of Species*. Only in 1871 did he publish his account of human origins, the *Descent of Man*, and by this time the general theory of evolution was beoming widely accepted. He also published numerous studies of botanical questions associated with the idea of evolution and a book on earthworms. He died after a long illness in 1882.

Peter J. Bowler is Professor of History of Science at The Queen's University of Belfast. He has published several studies of the history of evolutionism, notably *The Eclipse of Darwinism* (1983), *Theories of Human Evolution* (1986), *The Non-Darwinian Revolution* (1988), and *Life's Splendid Drama* (1996). He has also written non-specialist works on this and related topics including *Evolution: The History of an Idea* (1983) and *The Fontana History of the Environmental Sciences* (1992). He is currently working on a study of science and religion in early twentieth-century Britain.

8 Stokes and Kelvin, Cambridge and Glasgow, light and heat

DAVID B. WILSON

Victorian physics was largely Cambridge physics, and Cambridge physics was largely the creation of Sir George Gabriel Stokes (1819–1903) and William Thomson, Baron Kelvin of Largs (1824–1907). The Kelvin temperature scale, Stokes' parameters for polarised light, Stokes' law for a sphere moving through a viscous fluid: the language of today's physics quietly echoes yesterday's greatness. This chapter tries to recapture some of that greatness by placing the intertwined careers of Stokes and Kelvin within the broader story of Victorian physics and Victorian Cambridge University. Their collaboration and influence is symbolised by the fact that what is known as Stokes' theorem was suggested in a letter from Kelvin to Stokes and then set by Stokes as a question in a Cambridge examination taken by James Clerk Maxwell, generally acknowledged as *the* premier physicist of the century.

The careers of the two had close similarities. Both succeeded in Cambridge's Mathematical Tripos in the 1840s and obtained professorships that they would hold for more than five decades. Early research gained both election as fellows of the Royal Society of London in 1851. Thomson received a knighthood in 1867, Stokes was created a baronet in 1889, and Thomson was raised to the peerage in 1892, becoming Lord Kelvin. As elder statesmen of Victorian science, they occupied the presidency of the Royal Society for a decade, Stokes from 1885 to 1890 and Thomson from 1890 to 1895. In their seventies, each contributed original ideas to research on the new physics of X-rays, radioactivity, and electrons.

It was not obvious that the youthful George Stokes and William Thomson would become Cambridge minds, no matter how bright they struck others as boys. Both spent their earliest years in Ireland and had much closer family ties to Irish and Scottish universities than to

Cambridge. George grew up near Sligo, son of a rector in the Church of Ireland. His father and three older brothers had attended Trinity College, Dublin, though one brother had also gone on to Cambridge. William spent his earliest years in Belfast, son of a mathematics teacher who had graduated from Glasgow University. In 1832, William's father became Professor of Mathematics at Glasgow, and William matriculated there himself in 1834. At that time, George was at Rev. R. H. Wall's school in Dublin, which he attended from 1832 to 1835. In 1834, therefore, George and William, aged fifteen and ten, were lads of obvious intellect – but living in Ireland and Scotland. Very few of Cambridge's best students came from Ireland and Scotland.

Nevertheless, Cambridge beckoned. Of England's two older universities, Cambridge was more noted than Oxford – or the universities of Ireland and Scotland – for its mathematical curriculum. Students pursuing honours degrees at Cambridge took a comprehensive examination in mathematics and mathematical physics to conclude their studies. Those who passed this Mathematical Tripos were ranked from the very best (senior Wrangler) through the descending levels of Wranglers, senior optimes, and junior optimes. Strong finishes launched successful careers – in the Anglican Church, at Cambridge University and its several colleges, in secondary education, and the law. Preparation was intense, competition keen, and the consequences great. The Smith's prize examination followed the Tripos and attracted the strongest students. The highest accolade was to be senior Wrangler and first Smith's prizeman.

Mathematical genius and family guidance ultimately brought both boys to Cambridge. George's older brother had done well at Cambridge, placing as Sixteenth Wrangler (sixteenth out of ninety honours graduates) in 1828. Moreover, a classmate of his had become principal of Bristol College, a secondary school whose curriculum reflected Cambridge's. George attended Bristol College from 1835 to 1837, when he matriculated at Cambridge. William's guide was his father, who, as a Professor of Mathematics himself, knew Cambridge's reputation. William matriculated at Cambridge in 1841.

They met in 1843 while Thomson was an undergraduate. For a few years in the 1840s, for the only extended time in their lives, they lived in the same place, able to engage in daily discourse. Stokes had been senior Wrangler in 1841 and elected a fellow of his college, Pembroke. Just across Trumpington Street was Peterhouse, the college where Thomson was an undergraduate, highly favoured to be senior Wrangler in the Tripos of 1845. Stokes had prepared for the Tripos with the great coach of the day, William Hopkins, and Thomson was doing the same. Stokes had begun publishing his research in 1842 with a paper in the *Transactions of the Cambridge Philosophical Society*. Thomson had published his first paper in 1841 in the *Cambridge Mathematical Journal*, a paper he had written even before coming up to Cambridge. Between questions for examinations and subjects for research, these two young men had plenty to talk about, and they became fast friends in the process.

Their careers separated them from 1846 onward. Though Thomson finished as only Second Wrangler in 1845 (but first Smith's prizeman), he was elected a Fellow of Peterhouse and thus spent 1845–6 in Cambridge. In 1846 he gained the Professorship of Natural Philosophy at Glasgow, replacing William Meikleham who had recently died. Writing in support of Thomson's candidacy 'as a personal acquaintance of his', Stokes assured the electors that Thomson would 'ably discharge the duties of his office, and reflect credit on the University which has chosen him'. Stokes was right. Thomson held the chair until 1899, reforming the teaching of the subject and, in the process, bringing his own research to Glasgow undergraduates. Stokes remained in Cambridge, becoming Lucasian Professor of Mathematics in 1849. Holding his professorship until his death, Stokes established the Chair's importance to the Mathematical Tripos and supported reforms in Cambridge's physics education that eventually helped diminish his own role there.

Letters between the two comprise one of the great correspondences of Victorian science. For a few years after 1846, Thomson returned to Cambridge after completing the academic year at Glasgow.

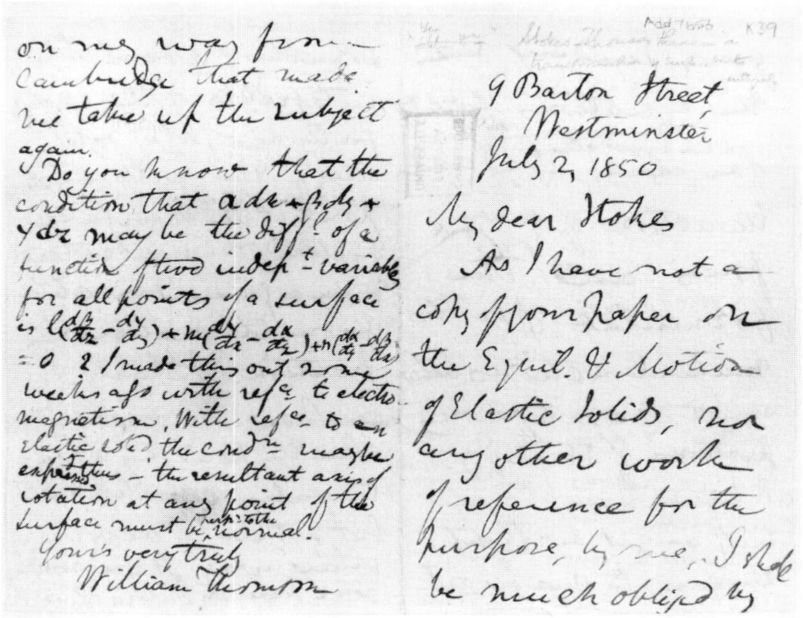
9 Barton Street
Westminster
July 2, 1850

My dear Stokes

As I have not a copy of your paper on the Equil. & Motion of Elastic Solids, nor any other work of reference for the purpose, by me, I shd be much obliged by

on my way from Cambridge that made me take up the subject again.

Do you know that the condition that $\alpha dx + \beta dy + \gamma dz$ may be the diffl. of a function of two indept. variables for all points of a surface is $l\left(\frac{d\beta}{dz} - \frac{d\gamma}{dy}\right) + m\left(\frac{d\gamma}{dx} - \frac{d\alpha}{dz}\right) + n\left(\frac{d\alpha}{dy} - \frac{d\beta}{dx}\right) = 0$? I made this out 2 or 3 weeks ago with refce to electro-magnetism. With refce to an elastic solid the condn may be expressed thus – the resultant axis of rotation at any point of the surface must be perp: to the normal.

Yours very truly
William Thomson

FIGURE 8.1 Kelvin to Stokes, 2 July 1850 stating Stokes' theorem. *Source:* Cambridge University Library, Stokes Collection, Add. MSS 7656, K39.

Telling his father in June of 1847 about his work with Stokes, Thomson wrote: 'I have been getting out various interesting pieces of work, along with Stokes, connected with some problems in electricity, fluid motion, &c., that I have been thinking on for years, and am now seeing my way better than I could ever have done by myself, or with any other person than Stokes.' Throughout their careers, they would see each other at scientific meetings and during Thomson's periodic returns to Cambridge, but they were to communicate mainly by letter. Overwhelmingly a professional correspondence between two eminent physicists, it also contains a personal side. They announced to each other their engagements in the 1850s. Stokes received several brief reports on Thomson's first wife's long illness and on Kelvin's neuralgia in old age. Stokes recommended that Kelvin try a particular 'neuralgic

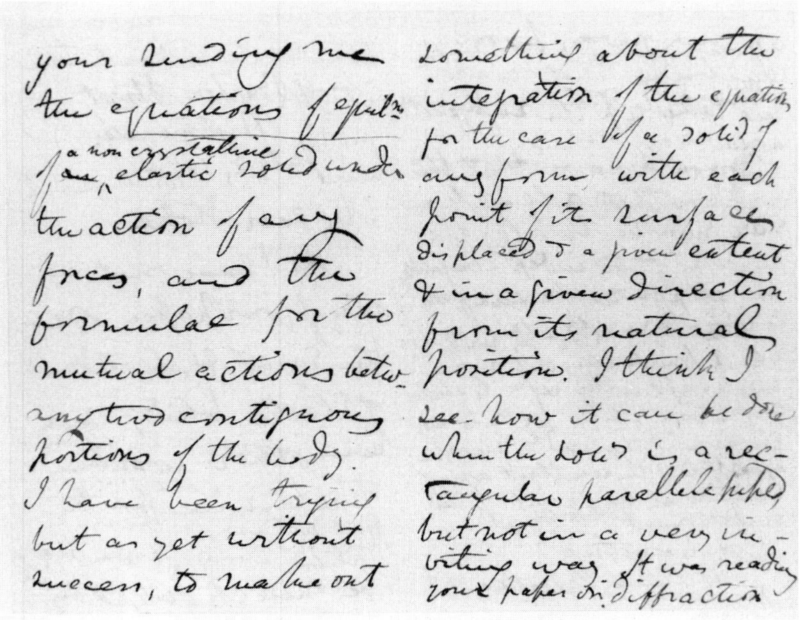

your sending me the equations of equil^{m} of a non crystalline elastic solid under the action of any forces, and the formulae for the mutual actions betw^{n} any two contiguous portions of the body. I have been trying but as yet without success, to make out something about the integration of the equations for the case of a solid of any form with each point of its surface displaced to a given extent & in a given direction from its natural position. I think I see how it can be done when the solid is a rectangular parallelepiped, but not in a very inviting way. It was reading your paper on diffraction

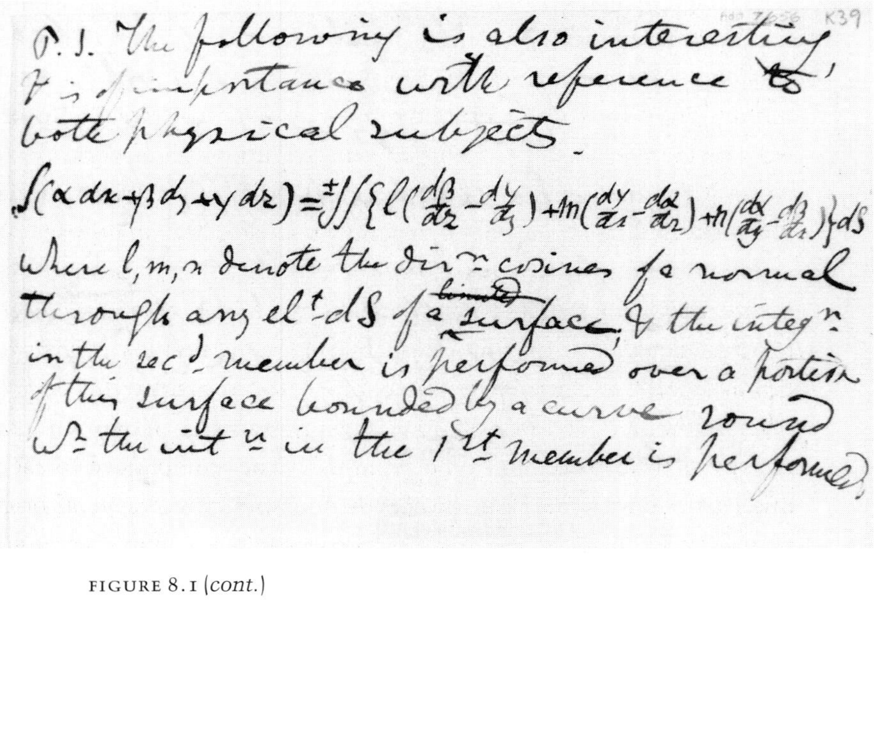

Add 7656 K39

P.S. The following is also interesting & is of importance with reference to both physical subjects.

$$\int(\alpha dx+\beta dy+\gamma dz)=\pm\iint\left\{l\left(\frac{d\beta}{dz}-\frac{d\gamma}{dy}\right)+m\left(\frac{d\gamma}{dx}-\frac{d\alpha}{dz}\right)+n\left(\frac{d\alpha}{dy}-\frac{d\beta}{dx}\right)\right\}dS$$

where l, m, n denote the dir^{n} cosines of a normal through any el^{t} dS of a ~~bounded~~ surface, & the integ^{n} in the sec^{d} member is performed over a portion of this surface bounded by a curve round wh^{h} the int^{n} in the 1st member is performed.

FIGURE 8.1 (*cont.*)

ointment' that Lady Stokes had found useful for rheumatic pain in her arm. Stokes had notes of sympathy following deaths of two children and his wife. In the 1890s, they congratulated each other on the jubilee celebrations of their professorships. In the memorial service for Stokes in Westminster Abbey, Kelvin declared that 'for 60 years of my life – from 1843 to 1903 – I looked up to Stokes as my teacher, guide, and friend'. This chapter concerns the scientific side of their lives.

The differences in their careers are even more illuminating than the similarities. The chapter's title alludes to these themes. The career of Stokes, the University of Cambridge, and the study of light – that was one natural cluster contrasting with a second; that of the career of Thomson, the University of Glasgow, and the study of heat. Their Cambridge minds had different slants, with different consequences for the history of physics, at Cambridge and elsewhere.

To explore these themes, we need first to realise that early Victorian physics could be generally divided into experimental and mathematical components. This distinction was not exactly the same as the modern one between experimental and theoretical physics. Both the experimental and mathematical portions involved theoretical conclusions, and mathematical physics incorporated experimental results. The difference was that the theories of mathematical physics were cast in the language of mathematics while those of experimental physics remained largely qualitative, and thus less advanced. This was not an absolute distinction, for experiments also could be precise and quantitative, and early nineteenth-century physicists, especially in France, had made progress in mathematising some of the 'experimental' areas of physics. But from the Cambridge perspective, the distinction held well enough to affect the content of the Mathematical Tripos.

Mathematical areas of physics corresponded closely to the achievements of Isaac Newton, Cambridge's Lucasian Professor in the late seventeenth century. Most conspicuous were his laws of motion and his mathematical theory of universal gravitation, set out in his *Principia* of 1687. Newton had pitted his theory against Descartes' qualitative theory of fluid vortices surrounding the sun and causing

planetary motions. As part of his attack on Descartes' theory, Newton developed the mathematical theory of fluid motion to reveal Descartes' mistakes. Newton's other great book, the *Opticks* of 1704, linked him to the geometry of light – to the mathematical analysis of the paths of light rays, whatever the physical nature of light. An exception was the mathematical wave theory of light, Newton having opposed earlier versions of it. Newton's well-founded objections had yielded, however, to the experimental research of Thomas Young around 1800 and the experimental mathematical work of Augustin Fresnel around 1820. Cambridge men thus generally embraced this new wave theory by the 1830s.

Experimental physics included mainly the subjects of heat, electricity, and magnetism. Though Newton had speculated in each area, none was a Newtonian subject like gravitational theory or geometrical optics. All three areas had been much cultivated during the century following Newton's death, of course, and associated with them by about 1830 were such familiar names as Lavoisier, Coulomb, Poisson, Oersted, Ampère, Fourier, and Carnot. Nevertheless, theories in these areas seemed less settled and not so satisfactorily mathematical as those in the realms of mathematical physics.

Cambridge's Mathematical Tripos had been reformed by Charles Babbage, William Whewell, and a few others around 1820, with the introduction of the continental version of the calculus. Along with pure mathematics, early Victorian Cambridge mathematical studies incorporated the physics of the time, in accordance with Cambridge educational principles. Mathematics being the exemplar of logical thought, the Mathematical Tripos grounded men in logical thinking, giving them a cast of mind suitable for engaging ideas in more cluttered areas. Including mathematical physics showed that clarity was possible outside mathematics itself. But the point was best made with established, not contentious, parts of physics. Hence, mechanics, hydrodynamics, gravitational theory, geometrical optics, and the wave theory of light dominated the Tripos' physical subjects. Whewell and others investigated the subjects of heat, electricity, and magnetism at

Cambridge in the 1820s and 1830s, but had little success inserting them into the Tripos. The Tripos existed within a supporting context of University Professors, College Fellows, the Cambridge Philosophical Society with its *Transactions*, and the *Cambridge Mathematical Journal*, first published in 1837. In the 1840s, these two journals published mainly articles by Wranglers on Tripos subjects. We can now place the research and teaching of Stokes and Thomson against this background.

Stokes' most intense period of research extended from the early 1840s through the early 1850s. Trying to make the physics of fluids more realistic, he took into account the internal viscosity that actual fluids possess. He examined a fluid's resistance to the motion of solids through it. He thus explained how such resistance influenced the swing of a pendulum and also how falling objects could reach a terminal velocity – the velocity at which any increased speed due to gravity was offset by the resistance of the surrounding fluid. The equations of motion for a viscous fluid turned out to be the same as those for an elastic solid. This conclusion, in turn, helped explain the physical nature of the luminiferous ether, the material medium that, according to the wave theory of light, transmitted the transverse waves constituting light. Invoking the principle of continuity, Stokes maintained that any material, no matter how solid or how fluid, would possess both solid and fluid properties. Fluid-like, the ether allowed planetary motions through it and, solid-like, supported transverse vibrations. With the same ether in mind, Stokes studied the optical phenomenon of diffraction, solving the problem of the exact orientation of light's transverse vibrations. In 1852, Stokes explained (and named) fluorescence, for which he won the Royal Society's Rumford medal.

Stokes' lectures as Lucasian Professor quickly became *de rigueur* for Cambridge's best students. Combining mathematical analysis and experimental demonstrations, he typically moved from fluids at rest to fluids in motion and then to sound waves as a transition to the wave theory of light. In 1852, Stokes showed his students experiments on fluorescence before he presented the results to the fellows of the Royal

Society. Though not required for the Tripos, his lectures became integral to it. During the first quarter century of his tenure, Cambridge averaged nearly forty Wranglers per year. Of those placing in the top ten, four of every five had attended Stokes' lectures. Of those finishing in the bottom half of the Wrangler lists, only one of five had been in Stokes' class. His professorial prominence later declined, partly due to Thomson.

Thomson, as already indicated, attended Glasgow University for several years before matriculating at Cambridge. In the Scottish tradition of higher education, Glasgow contrasted with Cambridge. A student's attention was divided more or less evenly among several subjects – Greek, Latin, moral philosophy, logic, natural philosophy, and possibly mathematics. Those were the subjects of the Arts Faculty. Chemistry was in the Medical Faculty and thus not a standard course for Arts students. Glasgow natural philosophy was more general and less mathematical than Cambridge mathematical physics. It included heat, electricity, and magnetism as well as the Newtonian subjects. The chemistry course was important for the study of heat, chemistry itself having been defined as 'the study of the effects of heat and mixture' by that greatest of Scottish chemists, Joseph Black, the discoverer of latent and specific heat in the eighteenth century. Glasgow Professors did provide the opportunity for the study of advanced mathematics and mathematical physics. Thomson's father, for example, encouraged advanced mathematical students to study the mathematics of modern mechanics and physical astronomy.

Thomson chose to study chemistry, as well as mathematics, natural philosophy, and the other subjects. Thomas Thomson, Professor of Chemistry, included heat and electricity in his course. He praised William's performance during the 1838–9 session and noted his work in the chemistry laboratory. The next year, William took the natural philosophy course, again covering the experimental areas of heat and electricity, as well as magnetism and the mathematical subjects of mechanics, astronomy, and light. In the following year of 1840–41, William's final year at Glasgow, he assisted the substitute

Natural Philosophy professor with experimental demonstrations for the lectures. Three reinforcing years thus afforded Thomson a knowledge of the likes of Oersted, Ampère, and Faraday that was *terra incognita* for Wranglers. He also had more experimental expertise than they. Like Wranglers, however, Thomson learned the significance of mathematical physics and studied it at an advanced level. Many of these strands came together in his mastery of Joseph Fourier's *Analytical Theory of Heat* (1822). Recommended by the Professor of Astronomy, J. P. Nichol, Fourier's sophisticated mathematical treatment influenced much of Thomson's later research. Indeed, Thomson's first paper, in 1841, defended Fourier's mathematics against criticism from the Edinburgh Professor of Mathematics, himself a former Wrangler.

Of Thomson's many papers on electricity and magnetism, let us mention three. In 1841, he explored similarities in the mathematical theories of static electricity and the flow of heat. An 1847 paper built on Michael Faraday's recent experimental discovery that a magnetic 'field' could rotate the plane of polarised light, a finding later known as the Faraday effect. Thomson envisioned various strains and motions within an elastic-solid medium as 'mechanical representations' of certain electrical, magnetic, and optical activities. It was intended as a mechanical analogue to the mechanical reality that underlay observable phenomena. In 1856, he presented a theory of spherical solids rotating within a fluid as an attempt to explain the Faraday effect and other phenomena. The rotating spheres would lend to the fluid an elasticity suitable for conveying transverse light waves. In the 1840s and 1850s, Thomson was thus seeking a unifying mechanical theory of various physical phenomena.

He contemplated heat, too. By the mid 1840s, he had been impressed not only by Fourier's book, but also by the research of Sadi Carnot. Carnot had analysed the function of a steam engine, relying on the theory that heat was conserved. Whatever exactly heat was, the total amount of heat remained the same, even though it could be shifted from place to place with temperatures of objects varying accordingly. In 1847, Thomson heard J.P. Joule lecture that heat was

the motion of material particles and that it was not conserved. He reported that his experiments showed that heat could be generated by large-scale motions of matter deteriorating into the motion of matter's smallest constituent particles. Evaluating Joule's discoveries in light of both Carnot's views and very recent research by Rudolf Clausius, Thomson published in 1851 what became known as the first and second laws of thermodynamics. He concluded that energy, not heat, was conserved, heat being merely one form of energy. Second, as energy changed from one form to another, it 'dissipated', or became less usable in an irreversible manner. Thomson also accepted Joule's 'dynamical' theory of heat, and its vision of moving material particles was another idea lying behind his 1856 paper on spheres rotating in a fluid.

As he was transforming physical theory, Thomson similarly altered the natural philosophy class at Glasgow. He continued his predecessor's practice of including a broad range of physical subjects, many of them now, however, informed by his own theories. He opened a separate laboratory where students could voluntarily do experimental work in addition to witnessing the demonstration experiments accompanying his lectures. Many of these students were effectively Thomson's assistants, performing experiments germane to his research – especially in areas of electrical technology, including the laying of the Atlantic Cable in the 1860s. Thomson also created a separate class for advanced mathematical physics, varying the course's topic.

Before returning to Cambridge and to Stokes' and Thomson's later careers, let us summarise the situation around 1860. In research and teaching, both Stokes and Thomson were emphasising a combined experimental-mathematical approach to physics. Each regarded physical nature as dynamical, with Stokes' idea of an elastic-solid medium holding great explanatory appeal for both. However, whereas Stokes had taken many of the subjects of the Mathematical Tripos to new levels of understanding, Thomson had probed the essence of the non-Tripos subjects of heat, electricity, and magnetism. Indeed, Thomson's deep conception of these subjects held the brilliant young Maxwell in

its sway. Influenced also by Stokes' treatment of elastic solids, Maxwell had already published a mathematical paper dealing with part of Faraday's work and was on the verge of publishing his revolutionary electromagnetic theory of light.

Sentiment grew in Cambridge during the 1860s to include these modern developments in physics education. As a result, the subjects of heat, electricity, and magnetism were added to the Mathematical Tripos in the early 1870s. Wranglers to be could read Thomson's *Reprint of Papers on Electrostatics and Magnetism* (1872) and Maxwell's *Treatise on Electricity and Magnetism* (1873). To provide mathematical students more experimental work, Cambridge created the Cavendish Professorship in experimental physics, hired Maxwell in 1871 to fill it, and opened the Cavendish Laboratory in 1874. Since 1851, the Natural Sciences Tripos had existed to promote the study of the natural sciences not represented by the Mathematical Tripos. As part of the reforms of the early 1870s, physics was included as a full-fledged subject – *experimental* physics, that is, because mathematical physics already had a tripos home. This persistent separation of experiment and mathematics was largely resolved around 1890 by J.J. Thomson, Cavendish Professor from 1884 to 1919. Physics in the Natural Sciences Tripos then became more mathematical than it had been, so that at the century's end prospective physicists (like C.T.R. Wilson) gravitated to the Cavendish Laboratory and the Natural Sciences Tripos. The Mathematical Tripos now tended to produce mathematicians (like G.H. Hardy) and astronomers (like A.S. Eddington). Just as physics education at Cambridge changed, so also did the places of Stokes and Thomson within Victorian physics.

Stokes' research became less intense, his teaching less influential. From the mid 1850s, Stokes was one of the great administrators of Victorian science, editing the Royal Society's *Transactions* for three decades and serving on committees in need of his technical expertise. Now his new ideas came primarily in response to research by others or in collaboration with others, especially less mathematically gifted physicists, or chemists. Most notable here was his work with William

Crookes from the 1870s on the newly produced phenomenon of cathode rays. The two established the British theory that the rays consisted of electrically charged particles, in disagreement with the German theory that they were ethereal waves. At Cambridge, physics education shifted increasingly to the expanding staff at the Cavendish and their experimental-mathematical treatment of the Thomsonian subjects of heat, electricity, and magnetism. In declining to apply for the Cavendish Professorship in 1884, Stokes wrote Thomson that he felt too old to take up new subjects and that the men under him at the Cavendish would know more than he did about those areas of physics. Continuing to lecture on the same subjects as before, Stokes found his course increasingly less relevant to students preparing for the Mathematical Tripos. Still, it was the Crookes–Stokes theory of cathode rays that lay behind J.J. Thomson's discovery of the electron in 1897, and Stokes' theory of X-rays as ethereal pulses did gain wide acceptance in the decade after their discovery in 1895.

In some ways, Thomson continued as before. More than Stokes, he maintained his own program of basic research in physics. His 1867 theory of vortex atoms continued his search for the dynamical essence of ether and matter. It described smoke-ring-like motions within a perfect fluid constituting both indivisible atoms of matter and an elastic-solid ether. The theory attracted followers, including J.J. Thomson who tried to develop it further in the 1880s. William Thomson's well-known *Baltimore Lectures* of 1884 was a wide-ranging and optimistic view of what he took to be well understood about the ether. Part of that understanding was that the ether should exhibit longitudinal (as well as transverse) waves. He had had this idea since mid century, and, in the 1890s, he at first thought that X-rays might be the sought-after longitudinal waves. More significant, his theoretical expectation of the existence of longitudinal waves was part of his continuing opposition to Maxwell's electromagnetic theory of light. Kelvin's active search for physical truth lasted into his final decade, when he wrote several papers offering a different theory of radioactivity and atomic structure than that being developed by J.J. Thomson and his student, Ernest Rutherford. Though

eventually eclipsed by the Thomson–Rutherford research program, Kelvin's theories were ingenious explanations of known data and point up some now easily overlooked ambiguities in early twentieth-century physics. Unlike Stokes at Cambridge, Kelvin remained at the centre of physics education at Glasgow until his retirement. Though he had a small staff, he was the dominant figure, still the only Professor of Natural Philosophy.

The strength of Cambridge University was perhaps the primary constant in this shifting story of Stokes, Kelvin, and Victorian physics. Stokes, Kelvin, and Maxwell had all been students there at mid century. Between 1870 and 1913, Britain educated ten physicists who won Nobel prizes. Nine had Cambridge degrees and the tenth was Rutherford's student at Manchester. Cambridge's constancy, however, should not veil the intriguing variables in Stokes' and Kelvin's respective careers. In investigating light and other Tripos subjects, Stokes' early research produced truly profound insights. In investigating heat and other non-Tripos subjects, Kelvin's early research changed physics. Though that change helped erode Stokes' position at Cambridge, it only enhanced Kelvin's dominance at Glasgow. Nevertheless, being at Cambridge, Stokes, not Kelvin, taught the élite of mid and late Victorian physics. Not teaching at Cambridge and disagreeing with Maxwell, J.J. Thomson, and Rutherford, Kelvin had not a single student of high eminence in physics. On a more positive, concluding note, both Stokes and Kelvin in late career, happily, dispelled any myth that original ideas are the exclusive province of young physicists.

Further reading

P. M. Harman. 1982. *Energy, Force, and Matter: The Conceptual Development of Nineteenth-Century Physics.* Cambridge: Cambridge University Press.

P. M. Harman (ed.). 1985. *Wranglers and Physicists: Studies on Cambridge Physics in the Nineteenth Century.* Manchester: Manchester University Press.

Joseph Larmor (ed.). 1907. *Memoir and Scientific Correspondence of the Late Sir George Gabriel Stokes.* 2 vols. Cambridge: Cambridge University Press.

Crosbie Smith and M. Norton Wise. 1989. *Energy and Empire: A Biographical Study of Lord Kelvin.* Cambridge: Cambridge University Press.

Silvanus P. Thompson. 1910. *The Life of William Thomson, Baron Kelvin of Largs*, 2 vols. London: Macmillan.

David B. Wilson. 1982. Experimentalists among the mathematicians: physics in the Cambridge Natural Sciences Tripos, 1851–1900. *Historical Studies in the Physical Sciences*, **12**, 325–71.

David B. Wilson. 1987. *Kelvin and Stokes: A Comparative Study in Victorian Physics*. Bristol: Adam Hilger.

David B. Wilson (ed.). 1990. *The Correspondence between Sir George Gabriel Stokes and Sir William Thomson, Baron Kelvin of Largs*, 2 vols. Cambridge: Cambridge University Press.

David B. Wilson. 1992. Scottish influences in British natural philosophy: Rise and decline, 1830–1910. In *Scottish Universities: Distinctiveness and Diversity*, eds. J.J. Carter and D.J. Withrington, pp.114–26. Edinburgh: John Donald.

M. Yamalidou. 1998. Molecular ideas in hydrodynamics. *Annals of Science*, **55**, 369–400.

William Thomson, 1st Baron Kelvin (1824–1907) studied science at Glasgow from the age of ten, and later was sent to Cambridge. He graduated when he was twenty-one and went to Paris to work on heat with Regnault. In 1846 he became Professor of Natural Philosophy at Glasgow, where he stayed for fifty-three years. In 1848 he proposed an 'absolute' scale of temperature now known as the Kelvin scale. He worked with Joule on the relation of heat and work (the first law of thermodynamics), and also with him found the Joule-Thomson effect. A pioneer of thermodynamics and electromagnetic theory, Thomson directed the first successful project for a transatlantic cable telegraph, which became operational in 1866, bringing him considerable personal wealth. In 1892 he was made a baron, and chose his title from a small river, the Kelvin, passing through the University.

George Gabriel Stokes (1819–1903) was educated in his native Ireland and at Cambridge. Stokes became Lucasian Professor at Cambridge in 1849 and in the next half century did much to rescue physics teaching there. He worked in most areas of theoretical and experimental physics except electricity. One of his enthusiasms was hydrodynamics and another was fluorescence, and both have laws named after him. In 1849 he pioneered studies of gravity variations over the Earth's surface; such methods are now used in stratigraphic studies to assist in oil prospecting.

David B Wilson is Professor of History, Mechanical Engineering, and Philosophy at Iowa State University. He is author of *Kelvin and Stokes: A Comparative Study in Victorian Physics* and editor of *The Correspondence between Sir George Cabriel*

Stokes and Sir William Thomson, Baron Kelvin of Largs. He received his doctorate in the history of science from The Johns Hopkins University. His research has focused on the history of physics, the history of science and religion, and science in Victorian Britain. He is currently studying natural philosophy during the Scottish Enlightenment.

9 James Clerk Maxwell
SIMON SCHAFFER

In Cambridge University's Cavendish Laboratory, alongside intricately impressive devices such as cloud chambers and mass spectroscopes, there is preserved an intriguing 'wheel of life'. A regularly slotted cylinder has a sequence of pictures carefully drawn on a removable strip of paper on its inner surface. As it spins vertically, persistence of vision gives a spectator peering through the slots the impression the figures are moving. This wheel is not on show merely because it was an important step towards stroboscopes and cinematography. It is there because this version was built in 1868 by a Glasgow instrument maker for James Clerk Maxwell (1831–1879), first head of the Cavendish Laboratory and supreme Victorian physicist. Maxwell put concave lenses of focal length equal to the wheel's diameter into the slits so every visible figure would be stationary. 'The unlearned pronounce it lively.' He designed fascinating images: boys play leapfrog, pine trees grow, acrobats leap, and fountains change colour. But he also showed cylinders in a resisting fluid, vibrating wires, and the behaviour of those interwoven vortex rings reckoned the basic constituents of matter by many nineteenth-century natural philosophers. He discussed his wheel of life with one natural philosopher, the Glasgow professor, William Thomson, then set its theory as a puzzle for the Cambridge Mathematics Tripos in January 1869. Maxwell had first played with zoetropes, to use their grander name, as a boy on his father's Glenlair estate in southwest Scotland in the 1830s. Later in life, he recaptured the spirit and hardware of childlike play and put it to effective use in the clarification of advanced physical theories for experts and the public. Few Victorians so well expressed the pleasure of scientific enquiry or drew so much profit from that pleasure.

Maxwell's repute principally relies on campaigns launched in the 1850s to explore and reorganise the principles of classical physics. He analysed colour vision and the stresses and strains of continuous media, demonstrated that electromagnetic action was propagated through such a medium, in which light itself travelled in transverse shear waves, and gave a compelling statistical account of gas kinetics and the behaviour of heat in large systems of interacting particles. Maxwell introduced a newfangled vector algebra to deal with some of these problems. He won important audiences and provided indispensable departure points for his successors in all the fields he explored. But he constantly teased out paradoxes which would undermine what had been established. Writings such as his magnificent *Treatise on Electricity and Magnetism* (1873) were provisional summaries and maps for further work, features which made them difficult handbooks for his new sciences. This sense of open dynamism is reinforced by Maxwell's comparatively early death. He would have been seventy-four when Albert Einstein made his decisive advance from Maxwell's principles of electromagnetism and optics; sixty-nine when Max Planck introduced the quantum of action to solve problems posed by Maxwellian thermodynamics; and barely fifty-seven when Heinrich Hertz provided decisive evidence of Maxwellian radiation as radio waves. When, in the 1870s, he contemplated puzzles with gaseous specific heats later to be addressed by quantum theory, Maxwell offered the Socratic consolation of a 'thoroughly conscious ignorance that is the prelude to every real advance in knowledge'. No conservative architect of high Victorian self-confidence, Maxwell was rather a provocative engineer of major changes in the edifice of traditional physical sciences.

Maxwell reached Peterhouse in 1850 after early training in Glenlair, then at Edinburgh Academy and, more happily, from 1847 at the Scottish capital's University, where his professors included the eminent logician William Hamilton and the expert thermal physicist and glaciologist James Forbes. Maxwell spent four years in Cambridge reading for the Mathematics Tripos, having quickly transferred to

Trinity and the stern tutelage in analysis and mechanics of William Hopkins, doyen of the university's fierce band of mathematics coaches. Belatedly elected to a Trinity fellowship in 1855, he left after a few months for the Natural Philosophy Chair at Marischal College Aberdeen, whose principal's daughter, the pious and devoted Katherine Dewar, he married in 1858. When his job was abolished in 1860, he tried for Forbes' place at Edinburgh before winning the Chair at King's College London. He retired in 1865 for residence in Glenlair and London. When the Cavendish Laboratory was established in 1871, Maxwell took the new Cambridge experimental physics professorship, held until his death from cancer in late 1879. His errant career contrasts with his two most important peers, Michael Faraday, who from the basement and podium of London's Royal Institution defined the behaviour of current-carrying wires and moving magnets, and William Thomson, Lord Kelvin, another Cambridge-trained mathematician who started the Glasgow University physics laboratory, thence promulgated a proudly Clydeside programme of precision measurement and lucrative investment in the techniques of electric telegraphy and steam engineering. Never a mere disciple, Maxwell's mercurial capacity cleverly to transgress fixed boundaries was much affected by the changing landscape of nineteenth-century physical sciences in the age of British industrial supremacy.

Natural philosophy, traditionally the study of all creation for evidence of wise design, was changed in the early nineteenth century by self-confident disciplines pursued in teaching laboratories, physical observatories, and public museums. Parisian institutions set the pace. There programmes developed by Pierre-Simon Laplace and his disciples in Napoleonic France applied celestial mechanics to other natural phenomena, understood as systems of particles acting instantly between their centres through empty space under mathematically precise force laws. Between the epochs of Waterloo and the European revolutions of 1848, this Laplacian programme was much transformed. Within France, its assumptions were challenged by Augustin Fresnel, who set out a wave theory of light, and Jean-Baptiste Fourier, whose

analytical mechanics of heat diffusion involved new methods of mathematising fluids' force and flux. In the German lands, the precision surveys which flourished in the wake of Laplacian celestial mechanics helped launch the work of Karl Friedrich Gauss, master of the theory of observational errors and, with his ally Wilhelm Weber, leader of a worldwide campaign in geomagnetism. The magnetic observatory they founded at Göttingen, and the contemporary optical institute started in Bavaria for Joseph von Fraunhofer, became headquarters of the production and management of fine new instrumentation for the physical sciences. Maxwell much admired the Gaussian programme: 'in every city of Europe you might see them, at certain stated times, sitting, each in his wooden shed, with his eye fixed at the telescope, his ear attentive to the clock, and his pencil recording in his notebook the instantaneous position of the suspended magnet. The scattered forces of science were converted into a regular army.' In emulation of Napoleonic scientific triumphs, German savants established institutes for such valuable enterprises as analytical chemistry and electro-technology. The institutes studied puzzles of electromagnetism, which they sought to model as a system of Laplacian forces, and of thermodynamics, which they began to understand through the interactions of large arrays of colliding particles. German protagonists of these approaches included some of Maxwell's most important colleagues from the 1850s – notably the mathematical physicist Rudolf Clausius, a clever investigator of gas dynamics, and the imposing Hermann Helmholtz, baptised H^2 in Maxwell's affectionate shorthand, promulgator of the principle of conservation of energy and eventually ennobled as head of the major new physics institute in imperial Berlin.

The British scientists' response to these striking European innovations was, as ever, eclectic and judicious. Cambridge's mathematicians reworked the algebraic analysis of Laplace's colleague Joseph-Louis Lagrange in their calculus textbooks and much studied the heat theory of Fourier and Fresnel's undulatory optics. But they sneered at undisciplined foreign passion for mere scientific novelty. In

1848 Cambridge mathematicians even agreed to expel heat theory and electromagnetism from the Mathematics Tripos, because these sciences were not yet securely cast in the deductive form fit for undergraduate consumption. When this ban was lifted in the 1870s and a university physics laboratory established, some dons tried vainly to persuade Thomson and then Helmholtz to head it before they turned to Maxwell. Suspicion of experimental and developing sciences remained. Cambridge scholars contrasted these novelties with the permanent truths of mathematical astronomy. As a young Cambridge graduate, Maxwell was invited by a college friend to produce a new English translation of Newton's *Principia*, the fundamental text of this model science. Laplace's development of Newtonian astronomy was widely followed. In the Smith's Prize which Maxwell jointly won at Cambridge in 1854, he was asked a typically Laplacian puzzle about the stability of the rings round Saturn. The next year, the dons again demanded an account of these strange celestial objects for the University's Adams prize, with Thomson as one of the judges. Maxwell's successful and lengthy answer showed they could not be solid structures, but must be a vast array of moving particles, which he compared with the cannonades then raging at Sebastopol.

By the time Maxwell scored these Cambridge triumphs his teachers, most of whose students were destined for the church, had sifted acceptable techniques in mathematical physics from its more subversive strains. There was a distrust of the materialist and even atheist implications of French astronomical sciences. Maxwell learnt the techniques and shared the distrust. He was never convinced that the force physics modelled on Laplacian astronomy was the right path for natural philosophy. Later in life, as public lecturer and physics editor for the *Encyclopaedia Britannica*, he taught that celestial mechanics was not as sure as it seemed. 'We can hardly say that there is even a beginning of a dynamical theory of the method by which bodies gravitate towards each other.' He guessed it had been an advantage that his hero Faraday's 'faith in the doctrine of attraction was dead', for the London experimental wizard was thus able to avoid the path of French

and German physicists too devoted to theories of point centres acting on each other at a distance, and could instead model electromagnetic action as tensed lines of force in space.

The stock-in-trade of Faraday and his colleagues, complex substances such as optical glass, high-pressure steam, or the gutta percha used to insulate telegraph cables, were understood by increasingly sophisticated hydrodynamics. Faraday was an expert in the manufacture of optical glass, and performed experiments demonstrating the rotation of the plane of polarised light in glass subjected to strong magnetic fields. Glass was like an elastic solid when subject to quick displacements, but a viscous liquid when slowly stressed. The dynamics which Maxwell practised described these strangely robust fluids using the partial differentials of physical properties across their unit segments. Fourier's analysis of heat flow helped, if stripped of unfortunate hypotheses about primordial molecules; so did the potential theory developed by Gauss to map geomagnetic forces. Once analysed as extended arrays of infinitesimal geometrical segments, the forces acting on each segment in some direction could be derived from total energy changes when a body moved in that direction. Cambridge mathematical teaching was crucial in making this new dynamics work. Hopkins trained Maxwell in differential analysis and the dynamics of continuous media, specifically applied to wave optics and fluid flow without worrying about the causation of gravity or the make-up of the luminiferous ether. Tripos questions were not merely ways of testing students' command of the doctrine; they could also be research topics which the examiners set to elicit novel mathematical work from their best candidates.

Maxwell revelled in the challenge, but suffered the formal discipline. In his attic rooms in Trinity in 1852, he drafted a poetic attack on Cambridge doctrine:

> that confounded hydrostatics
> sink it in the deepest sea! ...
> Such the eyes, through which all Nature

seems reduced to meaner stature.
If you had them, you would hate your
symbolising sense of sight.

Much later, he still had a jaundiced view of Tripos training: 'there are three ways of learning propositions – the heart, the head and the fingers; of these the fingers is the thing for examinations, requires great labour and constant practice, but dispenses with thought and anxiety together'. Maxwell's Cambridge career certainly had its anxieties. He found 'x and y innutritious, Greek and Latin indigestible and undergrads. nauseous. Is truth nowhere but in mathematics?' He took a lofty view of the gross excesses of May Week, suffered a profound spiritual crisis the summer before he sat the Tripos, and though very successful in the subsequent examination, yet failed immediately to impress some of his important superiors. His Edinburgh professor, Forbes, recognised Maxwell's remarkable abilities in mathematical physics, but was struck by the young man's 'exceeding uncouthness', so 'thought the Society and Drill of Cambridge the only chance of taming him'. Forbes' old friend, the highly conservative Master of Trinity William Whewell, agreed that Maxwell needed 'a little more culture of various kinds'. Even after Maxwell's successes in the Tripos and the Smith's Prize in 1854, Whewell's college initially refused the Scotsman a fellowship because his command of classical languages was weak and his mathematics lacked 'neatness and finish'.

Yet Maxwell also made much of the opportunities for intellectual work Cambridge connexions provided. He sent youthful papers in mechanics and optics to the progressive *Cambridge and Dublin Mathematics Journal*, presented the elite student club, the Apostles, with philosophic essays, attended lectures on such topics as the French Revolution and state-of-the-art engineering, contemplated producing a popular optics textbook, and, just after his Tripos success, was invited to give a brilliant paper on the mathematics of flexion to the prestigious Cambridge Philosophical Society. At the end of 1855 Maxwell gave the Society his first important paper on Faraday's electromagnetism. The

professor who invited him, George Stokes, was one of Maxwell's more important sources of inspiration at Cambridge. Maxwell first saw Stokes at a British Association meeting in Edinburgh in 1850, became an auditor of his superbly organised Cambridge optics lectures and then tried his ingenious spectroscopic experiments at home in his Glenlair attic. Stokes' research was indispensable for advances in hydrodynamics. His viscosity physics modelled that strange ether which must be rigid enough to transmit light waves, yet fluid enough to allow the free passage of bodies. Stokes' fluid dynamics helped make geodesy more accurate, transformed meteorology, was applied by Forbes to the motion of glaciers, and avidly absorbed by their greatest student, Maxwell. When Maxwell successfully sought the Aberdeen Chair in early 1856, both Forbes and Stokes backed him enthusiastically. Nor was his coach, Hopkins, in any doubt of Maxwell's stature: 'I have known no one who, at Mr Maxwell's age, has possessed the same amount of accurate knowledge in the higher departments of physical science. . . . it is impossible that he should not become (if his life be spared) one of the most distinguished men of science in this or any other country.'

Maxwell realised this promise in the later 1850s in work on heat engines and electromagnetic systems, topics absent from the Cambridge mathematics curriculum but of much interest in British engineering. The Scottish professors set the conservation of energy and the irreversible dissipation of available work as principles of their new sciences. Maxwell later explained that 'the first aim of science at this time should be to ascertain in what way particular forms of energy can be converted into each other'. He even supposed 'we can form a rough estimate of the efficiency of a man as a mere machine, and find that neither a perfect heat engine nor an electric engine could produce so much work and waste so little in heat'. Maxwell could imagine measuring human work and waste but denied the relevance of energetics to human nature. 'There is action and reaction between body and soul, but it is not of a kind in which energy passes from the one to the other.' Maxwell's keen sense of dynamics' importance, and his play with

apparently unchallengeable principles, let him see that modish energetics was by no means reducible to dynamics alone. The law which banned the spontaneous flow of heat from colder to warmer zones in a closed system, set forth by Thomson and Clausius, was a statistical, not a dynamical, principle. In a famous thought-experiment designed in 1867, then publicised in his handbook for 'working men', the *Theory of Heat* (1871), Maxwell imagined a 'small but lively being incapable of doing work' who could nevertheless violate the law. Thomson called Maxwell's infuriatingly agile being a 'demon', Maxwell thinking of it as a humble railway pointsman and mocking German physicists who tried vainly to turn this statistical regularity into a dynamical theorem.

The demon was turned on immediately relevant targets. The devout natural philosophy of Cambridge was increasingly hostile to the fashionable alliance of German energetics and Darwinian evolution peddled by public scientists such as John Tyndall, Faraday's Royal Institution successor, and his metropolitan allies. Maxwell had his own interests in public education in good sciences. As a Cambridge fellow he learnt from the Christian socialist Frederick Maurice that the colleges had originally been meant for 'work and study without retirement from the world'. For a decade from 1855 Maxwell taught at the working men's colleges which Maurice's movement supported. In an important lecture on Clydeside in honour of James Watt in 1873, he linked the social changes bred by steam technology with the new physics it had promoted: 'wrong opinions in dynamics are the causes of errors which run through every department of life. A great deal has yet to be done before even the whole of what is called the scientific world is thoroughly free of dynamical errors. And this will never be done merely by teaching everybody mathematics.' Maxwell's energetics studied mechanical systems which behaved erratically or capriciously, as did his early essay on the stability of Saturn's rings. At the end of his life he wondered whether the soul was 'like the engine driver, who does not draw the train himself, but, by means of certain valves, directs the course of the steam so as to drive the engine forward or backward or to stop it'. In early 1873 Maxwell read a paper to fellow dons about apparently continuous,

orderly, mathematical systems which nevertheless displayed sudden discontinuities. It was the solemn duty of public scientists to disseminate wider understanding of the 'singularities and instabilities' of physical systems, so that the bad effects of materialism and the reduction of humans to mere machines could be corrected and quashed. He recognised that 'such is the respect paid to science that the most absurd opinions may become current provided they are expressed in language the sound of which recalls some well-known scientific phrase'. He saw his new university laboratory as an institution for the correction of these errors.

Convinced of the urgent need to free the industrial and scholarly worlds of dynamical errors, Maxwell's cunning was often devoted to humble devices as tools for building better physics and public scientific culture. 'We may find illustrations of the highest doctrines of science in games and gymnastics', he announced in his inaugural lecture as Cavendish Professor. In studies of spinning coloured tops at the start of his career, or of William Crookes' puzzling radiometer at its end, he turned his ingenuity into new theories of light and heat. One highly fruitful case was the stress colours seen by polarised light, a phenomenon now used by engineers to check the robustness of complex model structures. These seemingly inconsequential colours started Maxwell on a path to statistical mechanics and dynamical electromagnetism. Certain crystals, such as calcite, divide light into two rays polarised at right angles along the directions of greatest and least elasticity. In 1847 Maxwell's uncle took the young scholar to visit an Edinburgh optician, William Nicol, who had designed a prism made of two calcite pieces cemented with Canada balsam, cut diagonally so that one polarised beam was transmitted and the other extinguished. Maxwell worked at Glenlair with his own polariscope, learning how to spot the colours seen under polarised light in glass sheets under stress. 'One has to be very cautious in sawing and polishing them for they are very brittle.' Maxwell derived algebraic expressions for lines of equal density in the glass matching those of equal heat as glass cooled. He quickly compared these with the lines of equal magnetic variation

mapped in Gauss' magnetic crusade. In return for some fine watercolours of these patterns, Maxwell got some Nicol prisms in summer 1849 which helped him extend his work to hot jelly poured between a paper cylinder and a cork, then twisted after cooling and viewed under polarised light. Laplacians supposed that compressibility and rigidity maintained a constant relation. But jelly was hard to compress yet easy to twist; cork was compressible but resisted shear. Well before reaching Cambridge, Maxwell linked these charming colours in stressed glasses and jellies to Stokes' work in hydrodynamics and the viscosity of fluids, then used them to criticise Laplacian models of molecular forces. He distinguished between pressures and their effects, the compressions; showed the proportionality of these factors' overall sums and net differences; then extended his mathematical principles of striations in stressed glass to rigid beams, to wax, resin, and gutta percha.

Maxwell published his essay on elasticity, with Forbes' desperate stylistic help, at the Royal Society of Edinburgh in 1850. He took his prized specimens of unannealed glass and water-colours to Cambridge. In his own family circle they were jokingly referred to as 'dirt'. He often returned to the lessons learnt from this dirt. As a Trinity Fellow in 1855 he taught undergraduates hydrostatics and optics. He analogised Faraday's lines of force, traced by iron filings, with streamlines in a flowing fluid, a further analogy to be obtained from 'a careful study of the laws of elastic solids and of the motion of viscous fluids'. Faraday's ingenious experiments on the magnetic rotation of the plane of polarised light suggested the relevance of Maxwell's youthful work on chromatic patterns of stress and strain, while Thomson's analogy of the laws of fluid flow to those of electrostatics and Stokes' analysis of the rigid but fluid ether proved helpful. Using a theorem discussed by Thomson with Stokes in 1850, then set by Stokes for the Smith's prize competition Maxwell jointly won at Cambridge in 1854, he could relate fluxes, the amount of fluid passing through an area, with forces, the intensity of fluid at that area. Puzzles of viscosity and elasticity were of equal salience when Maxwell first read Clausius's work on gas kinetics in 1859, after finishing his own great essay on Saturn's rings.

Clausius defined the mean free path particles would travel between collisions and assumed they would behave as if all moved at the same average velocity. Used to the problems of analysing large arrays of colliding particles, Maxwell countered that their speeds must be distributed in the same way as Gaussian errors. He decided to 'abstain from asking the molecules where they last started from, avoiding all personal enquiries which would only get me into trouble'. But maybe it was possible to get useful results by deriving the most probable behaviour of aggregate populations of molecules. If any fluid were a vast array of these elastic particles, its viscosity would depend on momentum exchanges in collisions. In a denser gas, bodies would collide more often with gas particles, but pairs of these particles would collide more often with each other. So, surprisingly, gas viscosity should be independent of gas pressure. This was just the sort of paradox Maxwell enjoyed, because it was tractable by dynamics and testable by precision measurement.

The same approach counted during the early 1860s in London, when Maxwell wondered what kind of fluid would account for the stresses he associated with Faraday's magnetic forces. By linking magnetic motions to the kinetic energy of fluid elements and electric displacements to the fluid's strains, he realised it could transmit shear waves at a speed almost exactly that of light. This he termed the Elecetromagnetic Theory of Light. His numbers were drawn from recent French astronomical observations and the precision measures of electromagnetic and electrostatic parameters at Weber's observatory in Goettingen. In alliance with Thomson, Maxwell soon joined a British Association campaign to apply Weber's programme to the needs of the British telegraphy industry. At King's College London his team used complex arrays of spinning current-carrying coils to get robust measures of electrical resistance embodying the energetics of electromagnetic systems. This work provided fundamental resources for the construction of his dynamical theory of the electromagnetic field in 1865, and for the first formulation of the relations later known as Maxwell's Equations. The enterprise also immediately suggested new

ways of investigating his puzzles with fluid viscosity and stress, made public by criticisms from Clausius. Maxwell had supposed, on the basis of his gas kinetics, that viscosity would be pressure independent, but proportional to the square root of temperature. The former notion was right. In 1865, experiments he and his wife performed at their London home with his rather accurate viscosity apparatus, spinning discs using a technique suggested by the spinning coils of the electrical resistance trials, showed however that viscosity varied linearly with absolute temperature. Never hesitant to revise firm views, Maxwell reorganised his gas theory. By analogy with his work on stress colours, he sought and found evidence of birefringence in moving fluids, revealing temporary double refraction lasting over intervals of mere fractions of a second. He could define the relaxation time, the period needed for stresses to be absorbed by fluids, which he used instead of the older notion of the mean free path. With a better model of the forces repelling gas particles from each other, Maxwell could derive known laws of partial pressures and gas diffusion. He could also use his new dynamical theory of electromagnetism to get a formula for birefringence without any questionable assumptions about the ether's mechanical behaviour. In works such as his *Theory of Heat* and his contribution to Longman's widely read *English Cyclopaedia* (1873), he summarised for public consumption the results of his long investigations into heat, light, and electromagnetism. All these phenomena were due to 'a medium truly material and capable of comparison with the kinds of matter with which we are more familiar'. His youthful play with polarising prisms, glass, and jelly stayed a formidable resource in the construction of his mature sciences.

Maxwell's career taught him ever more clearly that these playful enquiries must be turned into solidly compelling work. The machinery of the sciences was refined and made precise. Maxwell was an able designer of what he called experiments of illustration. He got an Aberdeen instrument maker to design a working model of the wavelike oscillations of Saturn's rings, and later commissioned a system of interacting gears to show how his electromagnetic theory of light

depended on the physics of coupled dynamic systems. Not everyone was convinced by these elegant mechanisms. His friend William Thomson, for example, judged that the 'so-called electromagnetic theory of light' was 'rather a backward step from an absolutely definite mechanical notion'. Thomson's worries also taught Maxwell the values of precision. Maxwell's evidence included the claim that Weber's electromagnetic and French astronomical estimates of light speed matched. But in 1876 Thomson still held that 'the result has to be much closer than has been shown by the experiments already made before the suggestion can be accepted'. Thomson was speaking at an important London conference and exhibition about the best instruments to use in science teaching. Maxwell wrote some of the programme notes for this show. He classified laboratory equipment as though it was so much factory stock. 'There must be a prime mover or driving power, and a train of mechanism to connect the prime mover with the body to be moved.' By making instruments look like thermodynamic machines, Maxwell helped link workshop with laboratory practices. He did the same in chapters of his *Treatise on Electricity and Magnetism*, where he outlined a hierarchy of tools, ranging from 'null or zero methods' such as the electroscope, through registration devices and scale readings, to 'instruments so constructed that they contain within themselves the means of independently determining the true values of quantities'.

Maxwell understood that these new techniques required new institutions, including the teaching laboratories set up in British universities in the wake of changes in electrotechnology and steam engineering. 'Electrical knowledge has acquired a commercial value', he told the readers of the new science magazine, *Nature*, in 1873, 'and must be supplied to the telegraphic world in whatever form it can be obtained'. His regime at the Cavendish Laboratory cultivated precision measurement of basic physical parameters, including electrical resistance. If the young laboratory could get competent workmen, reliable equipment and capable researchers, resources which Cambridge lacked, this campaign could help improve the data of his electromag-

netic theory. 'The desideratum', he joked, 'is to set a Don and a Freshman to observe and register (say) the vibrations of a magnet together, or the Don to turn a winch, and the Freshman to observe and govern him'. When several dons worried about the introduction of this kind of workshop culture into their cloistered world, Maxwell retorted that measurement of fundamental physical values was a proper part of spiritual life. In encyclopedia articles and lectures at the British Association during the 1870s he explained that 'those aspirations after accuracy in measurement, truth in statement and justice in action are ours because they are essential constituents of the image of Him who in the beginning created not only the heaven and the earth but the materials of which heaven and earth consist'.

Maxwell found the justification for precision sciences in his own religious faith, his keen understanding of physical theory and the pressing needs of contemporary technologies. Public sciences counted as part of his vision for the future. Maxwell praised scientists who had turned the 'analytical splendour' of abstruse dynamics into 'qualities which we now know to be capable of direct measurement and which we are beginning to be able to explain to persons not trained in high mathematics'. He mourned scientists' specialisation and wilful seclusion, for it made them seem like labourers rather than interpreters of the world. 'The very important part played by calculation in modern mathematics and physics has led to the development of the popular idea of a mathematician as a calculator, far more expert indeed than a banker's clerk', but someone whose work could be better performed by a machine. He urged the advantages of the new tools of vector algebra, which he used to such clever effect in his electromagnetic research, because they enable 'us to see the meaning of the question and its solution'. This was better than the mechanisms of mere analytical calculation. The same considerations applied to the future of his physics laboratory. 'The opinion seems to have got abroad that in a few years all the great physical constants will have been approximately estimated, and that the only occupation which will then be left to men of science will be to carry on these measurements to another place of decimals.'

Maxwell thought this opinion was wrong and dangerous. He never proclaimed the end of physics, but wanted the new Cavendish Laboratory to join in the enterprise of developing its future. He confidently expected that the wheel of life would keep on spinning.

Further reading

Lewis Campbell and William Garnett. 1884. *The Life of James Clerk Maxwell*. London: Macmillan [first edition 1882] second edition (with additional correspondence).

P.M. Harman (ed.). 1990, 1995. *The Scientific Letters and Papers of James Clerk Maxwell*, 2 vols. published to date, third volume to be published in 2002. Cambridge: Cambridge University Press.

C.W.F. Everitt. 1975. *James Clerk Maxwell: Physicist and Natural Philosopher*. New York: Charles Scribner's Sons.

P.M. Harman. 1998. *The Natural Philosophy of James Clerk Maxwell*. Cambridge: Cambridge University Press.

Daniel Siegel. 1991. *Innovation in Maxwell's Electromagnetic Theory*. Cambridge: Cambridge University Press.

Bruce Hunt. 1991. *The Maxwellians*. Ithaca, NY: Cornell University Press.

Crosbie Smith. 1998. *The Science of Energy: A Cultural History of Energy Physics in Victorian Britain*. London: Athlone Press.

Andrew Warwick. Forthcoming. *Masters of Theory: A Pedagogical History of Mathematical Physics at Cambridge University, 1760–1930*. Chicago, IL and London: Chicago University Press.

James Clerk Maxwell (1831–1879), natural philosopher, was born in Edinburgh on 13 June 1831, son of John Clerk Maxwell and Frances Cay, who died when Maxwell was eight. He entered Edinburgh Academy in 1840, and at the age of fifteen already published the first of a series of papers on the construction of geometric curves. From 1847 he studied at Edinburgh University, then in autumn 1850 entered Cambridge University, first at Peterhouse then soon at Trinity College, where he was coached for the Mathematics Tripos by William Hopkins. He graduated as second wrangler and joint first Smith's prizeman in 1854. As a new fellow of Trinity he published on colour perception and on Faraday's lines of electromagnetic force. Between 1856 and 1860 Maxwell was natural philosophy professor at Marischal College, Aberdeen. In 1858 he married the daughter of the College's principal, Katherine Dewar, with whom he shared a deep piety and a close personal affection. He completed a successful analysis of the stability of Saturn's rings, a problem set for the Adams prize of 1857,

demonstrating that they were composed of many small particles, an approach also used from 1860 to analyse heat and gas dynamics. He derived the gas laws from the effects of collisions between huge arrays of moving particles and identified the particles' velocity distribution in equilibrium with the normal error law. These studies laid the foundation of statistical thermodynamics. At Aberdeen much of his work was devoted to ingenious experiments on colour mixing and the phenomena of colour blindness, rewarded with the Royal Society's Rumford medal in 1860. In that year Maxwell moved to the natural philosophy chair at King's College London.

In London Maxwell produced crucial papers in 1861–2 and 1865 which offered a coherent mathematical account of the electromagnetic field, applied Lagrangean analysis to the behaviour of electromagnetic energy stored in a space-filling ether, and demonstrated that transverse waves propagated in this ether would have the same speed as light. Subsequently he set out the 'general equations of the electromagnetic field', the fore-runners of the four Maxwell Equations. In 1862 Maxwell joined a new British Association committee to establish exact standards in electromagnetic measurement, notably a precise value for electrical resistance, a quantity of importance for practical telegraphy. He directed experiments during the 1860s to establish this value and thus also provide numerical evidence for his theoretical calculations of the velocity of electromagnetic waves. He resigned from the King's College professorship in 1865, then spent his time in London and his Scottish estate, Glenlair.

Having served as an examiner for the Mathematical Tripos, in 1871 Maxwell became Cambridge's first Professor of Experimental Physics and set about constructing the university's new physics laboratory. He edited the electrical papers of Henry Cavendish, distinguished kinsman of the laboratory's patron, the Duke of Devonshire; wrote important textbooks in mechanics, thermodynamics, and electricity; and published his great *Treatise on Electricity and Magnetism* (1873), in which he presented the mathematical and practical basis of the dynamical theory of the electromagnetic field. He lectured for the Mathematical and Natural Sciences Triposes, and, though by no means an active supervisor of experimental work, inspired the first generations of Cambridge experimental physics students trained in the laboratory. Maxwell also presented a series of widely read lectures and encyclopaedia entries on major themes of atomic and molecular physics. He used these new platforms to contest what he saw as the more dangerous implications of radical materialism, determinism, and evolutionism espoused in the name of the new sciences. He died in Cambridge on 5 November 1879.

Simon Schaffer is Reader in History and Philosophy of Science at Cambridge University and Fellow of Darwin College. He was trained in history and philosophy

of science at Cambridge and Harvard, and has published on the social history of natural philosophy, astronomy and experimental physics. He has co-edited books on Robert Hooke and on William Whewell, on the history and sociology of experiment, and on the sciences of enlightened Europe. He is the coauthor of *Leviathan and the Air Pump* (Princeton, NJ, 1985).

10 **The duo from Trinity: A.N. Whitehead and Bertrand Russell on the foundations of mathematics, 1895–1925**
IVOR GRATTAN-GUINNESS

INITIAL ENTRIES INTO FOUNDATIONAL STUDIES

Mathematics occupied a controversial place in Cambridge in the late nineteenth-century: the Tripos was being roundly criticised as a mere set of skills, and yet it must have helped the university to gain a high reputation in applied mathematics. Alfred North Whitehead (1861–1947) started off in this branch after graduating from Trinity in 1884, being quickly elected to a college Fellowship with a dissertation on Maxwell's theory of electromagnetism. Further work drew him to the algebraic methods of the German mathematician Hermann Grassmann, which he popularised in a large book called *A Treatise on Universal Algebra, with Applications* (1898). The title was a misnomer, in that no one algebra was presented but instead a range of them, including also George Boole's algebra of logic.

Pure mathematics at Cambridge was rather boring, with excessive emphasis laid upon linear algebras due to Professor Arthur Cayley, and rather routine treatments of the calculus and analysis. Bertrand Russell (1872–1970) took the Mathematics Tripos from 1890 to 1893 (with Whitehead as one of his tutors), but then abandoned the subject in disgust and moved over to philosophy. He united these two trainings in an attempt to find a foundations for mathematics, starting with a Trinity Fellowship dissertation in 1895 which he revised into the book *An Essay on the Foundations of Geometry* (1897). His philosophical training lay in the neo-Hegelian tradition then dominant, which he exercised with skill; but the results for mathematics were

not satisfactory. So when his Trinity friend G.E. Moore (1873–1958), who was then working largely on ethical theory, revolted against the tradition in 1899 and put forward a strongly empiricist alternative, Russell joined him. Thereafter he construed both logic and mathematics in that spirit as much as possible.

THE IMPACT OF PEANO AND CANTOR

During the late 1890s Russell tried various approaches to the foundations of mathematics, including using aspects of Georg Cantor's set theory and Whitehead's book; by 1900 he had a book nearly complete on 'Principles of mathematics'. Then he and Whitehead went to Paris late in July 1900 for the International Congress of Philosophy. This was a turning point of the philosophical careers for both men. The decisive participant was the Italian mathematician Giuseppe Peano (1858–1932), who had built up a school at Turin bent on raising the level of rigour in mathematics by formalising the symbolism of both logical and mathematical notions. A whole morning was devoted to this enterprise, and both Russell and Whitehead realised its superiority over other foundational methods.

Peano was much influenced by the programme of rigour in mathematical analysis that Karl Weierstrass had advocated in his lectures at Berlin University especially from the 1860s (correctly formed definitions, full details of working, and so on), and decided that ordinary language could be too ambiguous to supply the care required. So he introduced much more symbolism, not only for the mathematics involved but also the 'mathematical logic' (his name) used in forming and handling propositions and making deductions. This logic comprised the calculi not only of propositions but also of 'propositional functions', such as 'x is a man' which take a true or false proposition as value when x is given an appropriate value. Set theory took a *central* role in both this logic and the mathematics expressed in its terms.

In his autobiography, Russell tells us that he received all of Peano's publications in 1900 'and immediately read them all', and then wrote an entire book during the rest of the year. Luckily he kept the

manuscript, so that we can see that the story is absurdly wrong; the writing and re-writing lasted until 1902.

After studying the 'Peanists' (as they were known), Russell quickly added a logic of relations (such as 'x is the brother of y') which they had not developed. Then he drafted four of the seven parts of the new book in 1900, treating integers, continuous quantities (which he reduced to integers by defining irrational numbers in terms of rational numbers), and various aspects of set theory. However, he did not yet have any clear philosophical position about the foundations; this came to him early in 1901. In expressing mathematical theories in logical form, the Peanists distinguished mathematical from logical notions; however, the division line between the two was not clear, especially concerning set theory, which turned up on both sides. Russell decided that such a division did not exist: that his logic (with relations) could deliver *all* 'pure mathematics' (a curious use of the adjective), not only the reasoning required but also its objects in the form of implications in which the components of the propositions involved were drawn from logic. This position has become known as 'logicism', a name coined (in this sense) by Rudolf Carnap in the late 1920s; it will be used below.

RUSSELL'S FIRST LOGICISM: *THE PRINCIPLES OF MATHEMATICS*

The mathematical base of logicism lay in Russell's new definitions of integers as classes of 'similar' classes, that is, those in one-to-one correspondence with each other. The definition of 0 was as the class containing the empty class, 1 was the class of classes similar to the one containing 0, 2 the class of classes similar to the one containing 0 and 1, and so on – including the theory of 'transfinitely' large numbers which Cantor had developed. One notable consequence of this theory was clearly to distinguish apart 0, the empty class and literally nothing, a tri-distinction which had plagued mathematics and philosophy for centuries and where even Cantor had been unsatisfactory.

Upon this basis Russell then defined rational numbers, then

irrationals, and thus the real line. Then the calculus could be constructed from functions between continuous variables and the theory of limits; then geometry could be formed in terms of the continuity of lines, planes, and volumes, and relationships between objects in them; and finally (some) mechanics was captured by adding in the continuity of space and time.

This was the vision sketched out in detail in the book, which appeared in 1903 from Cambridge University Press as *The Principles of Mathematics*. However, in it Russell also reported bad news: in manipulating classes, he had discovered that the logic admitted contradiction. Working a variant of a (legitimate) proof method due to Cantor, he considered the class of all classes which do not belong to themselves, and drew the logically distressing conclusion that it belonged to itself *if and only if* it did not do so. Here was *a double* contradiction, not just the single contradiction used in mathematics in contexts such as proof by reduction to the absurd. He added an appendix to the book proposing a solution, but soon afterwards he realised that it failed.

In developing his logic, Russell noticed its bearing upon language and the possible referents of words. He became especially concerned in his book with six little ones: '*all, every, any, a, some,* and *the*'. The first and fifth were especially important, for they pertained to quantification theory (not his invention), where a proposition was formed from a propositional function by saying 'for all x...' and 'there exists an x such that ...' and the variable x became 'apparent' because its values had been 'absorbed'. He tried to use a consistent vocabulary for notions within and without logic: words such as 'proposition', 'propositional function', 'variable', 'term', 'entity', and 'concept' denoted *extra*-linguistic notions; pieces of language indicating them included 'sentence', 'symbol', 'letter', and 'proper name'. A word 'indicated' a concept which (might) 'denote' a term.

Russell had a small circle around him at Trinity. Above all, Whitehead was quite interested, and after Paris began to rework parts of Cantor's theory of infinitely large numbers in algebraic form. He also found Russell's post-Peano programme more clearly focused than his

own, and gradually their consultations turned into a formal collaboration to elaborate Russell's vision. Further, the mathematician G.H. Hardy (1877–1947) was then just starting his career with some papers in set theory; and a former student, Philip Jourdain (1879–1919), was also working on set theory and logic and came later to function more effectively as their historian. In addition, Moore (who moved to Edinburgh in 1904 for five years) was sympathetic to the enterprise.

After sending his book to the Press, Russell began to read the work of the German mathematician Gottlob Frege (1848–1925), who had anticipated both his logicistic thesis (though asserted only of arithmetic) and some uses of mathematical logic. So, while his book was on proof, he used Frege to change a few passages in the main text and added an appendix reviewing Frege's achievements. However, he had already formed versions of many features of Frege's theory; some philosophers greatly exaggerate the extent of its influence on Russell, both then and later.

PARADOXES AND INDECISION

Russell realised that his paradox was very serious, not merely like a neo-Hegelian class of opposites. So he collected other paradoxes, or at least gave paradoxical status to certain results known previously. Two especially important ones concerned the largest possible infinite cardinal and ordinal numbers; assumption of either of their existences also led to contradictions. The classical Greek paradox 'this proposition is false' was thrown in; and also new paradoxes of naming, such as 'the smallest number which cannot be defined in less than 23 syllables' and yet has just been done so. He hoped to find a common solution to all these paradoxes, not just a means of avoiding them. From 1903 to 1907 he tried all sorts of means; however, some of them let paradoxes in by another route, while others also excluded legitimate parts of mathematics and set theory.

A major task for logicism was to express in terms of propositional functions the 'denoting functions', especially ordinary mathematical functions such as x^2. One special difficulty in denoting, not confined to

mathematics, was with propositions like 'the present King of France is bald', since no person had the honour of holding this post at the time. Given that such propositions seemed to be meaningful, were they true or false? In 1905 Russell made a big advance in laying down conditions for the existence of 'definite descriptions' (to use his later name) of this kind within a proposition: that there should be one and only one entity involved, and it should indeed have the property required; otherwise the proposition was false. These criteria were just like those long known in mathematics for a mathematical function to be single-valued; indeed, Peano had stated them in that context in 1897, although Russell had presumably forgotten the text when proposing his more general definition. The theory also exhibited his Moore-style parsimony: a denoting function was used in the *context* of a proposition asserting some property (such as baldness), not in isolation.

The theory had a further consequence. During 1905 and 1906 Russell abandoned the idea of variables at all and worked just with the notion of 'substitution', where he took a proposition such as 'Socrates is a man' and substituted 'Plato' for 'Socrates' to obtain the (in the sense of a definite description) proposition 'Plato is a man'. However, he eventually found that this theory admitted a paradox, and anyway it offended his empiricist inclinations by requiring truth-values of propositions to be objects.

Meanwhile, Whitehead was thinking about geometry and space. In 1905 he published a long paper with the Royal Society using these logical techniques to present a construction of 'the material world' of points, instants and particles. He also contributed two volumes to the new series of Cambridge Mathematical Tracts, on projective geometry (1906) and descriptive geometry (1907).

PREPARING *PRINCIPIA MATHEMATICA*

During 1906 Russell abandoned the substitutional theory and went back to variables and propositional functions; and soon they thought they had a clear enough vision of their programme for a book to be prepared. By this time it was no longer a successor to Russell's *Principles*

but an independent work called *Principia Mathematica*; a certain book by their Trinity predecessor Isaac Newton may have been a model for this title, and probably also Moore's *Principia Ethica* (1903).

From early in 1907 until late in 1909, the two men worked out the book. Each man took first responsibility for a section or part, and then had his material scrutinised by the other. They were not often together, for in 1905 Russell had built himself a house at Bagley Wood in Oxford while Whitehead was living at Grantchester near Cambridge. In the end, Russell actually wrote out the entire text, though much of the content was Whitehead's. Several parts were read by another former student, Ralph Hawtrey (1879–1974), just then launching his career as an economist.

They already had the agreement of Cambridge University Press to publish the book; but when the crates arrived, the accountants panicked. So in 1910 the authors successfully obtained £200 from the Royal Society to help defray the costs of printing. Eventually three volumes appeared in 1910, 1912, and 1913.

THE SECOND LOGICISM: *PRINCIPIA MATHEMATICA*
The text began with an account of the calculi of propositions and propositional functions with set theory, including the theory of definite descriptions and the definability of mathematical functions. In line with parsimony, classes were defined in terms of propositional functions, again contextually. Then they laid out in enormous detail the theory of both finite and transfinite arithmetic, starting out from Russell's definitions of integers; figure 10.1 shows a vital stage in the development.

The means finally decided to Solve the paradoxes was based upon a 'vicious circle principle', which stipulated that, 'All that contains an apparent variable must not be one of possible values of that variable.' Propositional functions were classified into 'orders' by the quantified variables which they contained; and within each order the remaining variables further specified a 'type'. The initial type was the class of individuals; then came the types of classes of them, classes of classes of

*110·632. $\vdash : \mu \in \mathrm{NC} . \supset . \mu +_c 1 = \hat{\xi}\{(\exists y) . y \in \xi . \xi - \iota`y \in \mathrm{sm}``\mu\}$
Dem.
$\vdash . *110\cdot631 . *51\cdot211\cdot22 . \supset$
$\vdash : \mathrm{Hp} . \supset . \mu +_c 1 = \hat{\xi}\{(\exists \gamma, y) . \gamma \in \mathrm{sm}``\mu . y \in \xi . \gamma = \xi - \iota`y\}$
[*13·195] $= \hat{\xi}\{(\exists y) . y \in \xi . \xi - \iota`y \in \mathrm{sm}``\mu\} : \supset \vdash . \mathrm{Prop}$

*110·64. $\vdash . 0 +_c 0 = 0$ [*110·62]
*110·641. $\vdash . 1 +_c 0 = 0 +_c 1 = 1$ [*110·51·61 . *101·2]
*110·642. $\vdash . 2 +_c 0 = 0 +_c 2 = 2$ [*110·51·61 . *101·31]

*110·643. $\vdash . 1 +_c 1 = 2$
Dem.
$\vdash . *110\cdot632 . *101\cdot21\cdot28 . \supset$
$\vdash . 1 +_c 1 = \hat{\xi}\{(\exists y) . y \in \xi . \xi - \iota`y \in 1\}$
[*54·3] $= 2 . \supset \vdash . \mathrm{Prop}$

The above proposition is occasionally useful. It is used at least three times, in *113·66 and *120·123·472.

FIGURE 10.1 A striking passage from *Principia Mathematica* (volume 2, pages 82–83 of the second edition), where some major theorems in arithmetic are proved and their importance modestly noted.

them, and so on finitely up. The same division was applied to pairs, triples,... of individuals, so partitioning relations in a similar way. Any object in this hierarchy could only belong to an object in the immediately superior level; thus (non-)self membership was forbidden, and so Russell's and other paradoxes avoided. In addition, propositions were stratified by their truth level, so that, for example, the falsehood of 'all propositions are true' was located at the next level; hence the liar and naming paradoxes were also evaded.

However, several deficiencies attended this theory. Firstly, the stratification into types meant that the various kinds of number were all located over the place, rendering arithmetic impossible; for example, 2, 1/2, and $\sqrt{2}$ could not be handled together. So with reluctance and without philosophical justification the authors proposed an 'axiom of reducibility', which asserted that to each propositional function in any

type there corresponded a logically equivalent one free of quantifiers, so that arithmetic and thereby mathematics could be constructed in all types.

Secondly, in order to meet mathematical needs, especially Cantor's theory of transfinite numbers, they required an *infinitude* of individuals in the bottom type. Russell's empiricism caused philosophical difficulties here. Individuals are hardly logical objects, but they were forbidden from being abstract structureless ones; so they had to be physical objects, thus making logic *a posteriori*. To minimise this unwelcome feature he imposed a rule that whenever possible only one individual should be assumed. However, when developing the theory of cardinal numbers Whitehead had forgotten to use this rule, and Russell had failed to notice; the error was spotted only when the second volume was in proof, and many months were lost while Whitehead effected an (obscure) repair. This is why the second volume appeared two years after the first.

A third difficulty concerned set theory itself. In 1904, Russell had realised that the technique of independently selecting members from an infinite collection of classes could not itself be proved from the principles of set theory; so he had to invoke a new axiom for the purpose, which he called 'multiplicative axiom' because of the particular context in which it came to light. Shortly afterwards, the German mathematician Ernst Zermelo found the need for the same sort of axiom in a different context of set theory, later calling it 'the axiom of choice'. Much controversy ensued among mathematicians and philosophers about the legitimacy of these axioms, and also the places where they could not be avoided. But there was an additional problem for the logicists: they always worked with *finitely* long propositions in their logic, whereas the axiom admits an infinitude of independent selections to be made.

In addition, the form of the logicism was unclear. The introduction to the book did not explicitly state the thesis at all, and the text used both the implicational version of *The Principles* and also an inferential

form, where from an asserted premise a conclusion would also be asserted. The scope of mathematics to be logicised was also not discussed: while they presented Cantorian mathematics in superb detail and laid the groundwork for aspects of geometry, they made no effort to cover the rest of Russell's earlier book – no calculus, no mechanics, and no explanation.

AFTERMATH

During the years of preparation of *Principia Mathematica*, Russell and especially Whitehead gave lecture courses at Cambridge on these topics from time to time; but none of the papers which they produced over the years appeared in a Cambridge periodical. However, for the Fifth International Congress of Mathematicians, held at Cambridge in 1912, Russell organised a session on the 'philosophy and history' of mathematics, although he did not give a paper.

In 1910 Whitehead moved to London, securing in 1912 a readership in applied mathematics at University College London and moving two years later to a chair at Imperial College, which he held until 1924. In the 1910s he was still working on the fourth volume of *Principia Mathematica*, which would present several aspects of geometry, including his construction of points. However, he also devoted much time to mathematics education and relativity theory; then he abandoned the volume, seemingly in reaction to the loss of a son during the Great War in 1918, and none of the manuscript survives. In 1924 he moved from London to Harvard University, where he devoted himself to philosophy of kinds in which logicism was absent, although he sketched out an alternative foundation for it in 1934.

The situation with Russell was quite different. Elected in 1910 to a lectureship at Trinity, in effect to replace Whitehead in foundational mathematics, he used his logic, especially the logic of relations, as an inspiration and even technical resource to develop an empiricist epistemology in terms of the reception and organisation of sense-data. An outline was given in his short book *The Problems of Philosophy* (1912); and in much more detail in *Our Knowledge of the External World as a*

Field for Scientific Method in Philosophy (1914: again, the account of the one-go writing given in his autobiography is absurdly mistaken). This book was based on a course delivered at Harvard, where he also gave one on *Principia Mathematica*: they helped to stimulate considerable interest in Russellian epistemology and logic(ism) in the USA.

In between these two books Russell was visited at Cambridge by a precocious American teenager, Norbert Wiener (1894–1964), already holder of a Ph.D. at Harvard University with a comparison of the logic of relations in *Principia Mathematica* with that of the algebraic logician Ernst Schröder. While Wiener published none of this study, the contact with Russell led him to write some papers on set theory and logic which were accepted via Hardy by the Cambridge Philosophical Society. The best-known one, inspired by the axiom of reducibility, showed that ordered pairs, triples, . . . could all be defined in terms of classes alone.

At this time another visitor arrived, from Manchester University: the Austrian engineer Ludwig Wittgenstein (1889–1951). They created between them the truth-table method of representing logical connectives as functions of the truth values of the connected propositions. But Wittgenstein also criticised an epistemological book that Russell was preparing so severely that Russell abandoned it, though some chapters appeared in the American journal *The Monist*, of which Jourdain was the European editor. *Our Knowledge* was the replacement book.

Then Russell's philosophical career was largely overtaken by the War, where his active pacifism (as it were) led in 1916 to his dismissal from the Trinity lectureship. After the War he resumed his epistmological programme with a book on psychology and lecture courses in London. He married for the second time in 1921, and lived mostly outside Cambridge. He spent a lot of his time on educational tasks, and on money-raising books; philosophical work occurred more fitfully, and research in logic stopped almost completely. But he was also involved in the publication of Wittgenstein's *Tractatus* (1921, 1922); and then with Frank Ramsey (1903–1930), who both helped him

proof-read a second edition of *Principia Mathematica* (1925–1927, for some reason the first two volumes were re-set) and with revision of the foundations of Russell's logicism. Partly under the influence of Wittgenstein (who rejected logicism), both Russell and Ramsey went for a more extensional conception of logic and mathematics; that is, where classes and thus propositional functions were regarded as composed of their members rather than being defined by some property.

Apart from these students, and interested spectators, such as Moore and Hardy, Cambridge took little part in developing either mathematical logic or the logicist programme. However, John Maynard Keynes was influenced by its general tenor when writing his *A Treatise on Probability* (1921), and soon afterwards the young topologist Max Newman (1897–1984) examined some epistemological issues. An unusual manifestation was evident in *The Calculus of Variants* (1927) by W. W. Greg (1875–1959), who had been the librarian at Trinity College when Russell and Whitehead were there. In this 'essay on textual criticism' Greg used the logic of relations to confect a symbolic representation of the relationships between an original text and its various later versions, and thereby became a founder of the notion of the copy-text.

Around 1930 the main interest in Britain switched to the philosopher Susan Stebbing (1885–1943), a Girton graduate but then at the University of London, who popularised and furthered Russell's cause, especially in her *A Modern Introduction to Logic* (1930). Both Russell's logic and epistemology played a role in the gradual rise of the (unhappily named) 'analytic philosophy', which came to supremacy in the Anglo-Saxon world. One of her connected interests lay in the work of the Vienna Circle of philosophers, several of whom took much note of Russell's philosophy themselves. The epistemology enchanted their leader, Moritz Schlick; both the logic and epistemology strongly influenced Carnap; and the logic affected Kurt Gödel, who showed in a famous theorem of 1931 that, in systems like *Principia Mathematica*, true propositions could be stated but not proved, so that an assumption that Whitehead and Russell had made in their conception of logicism was untenable.

Further reading

A. Garciadiego. 1992. *Bertrand Russell and the Origins of the Set-theoretic 'Paradoxes'*. Basel: Birkhäuser.

I. Grattan-Guinness. 1997. How did Russell write *The Principles of Mathematics* (1903)? *Russell*, n.s. **16**, 101–127.

I. Grattan-Guinness. 2000. *The Search for Mathematical Roots, 1870–1940. Logics, Set Theories and the Foundations of Mathematics from Cantor through Russell to Gödel.* Princeton, NJ: Princeton University Press.

N. Griffin. 1994. *Russell's Idealist Apprenticeship*. Oxford: Clarendon Press.

P. Hager. 1994. *Continuity and Change in the Development of Russell's Philosophy*, Dordrecht: Kluwer.

G. Landini. 1998. *Russell's Hidden Substitutional Theory*. New York: Oxford University Press.

V. Lowe. 1985, 1990. *Alfred North Whitehead. The Man and His Work*, 2 vols. (vol. 2 ed. J.B. Schneewind). Baltimore: Johns Hopkins University Press.

F.A. Rodriguez-Consuegra. 1991. *The Mathematical Philosophy of Bertrand Russell: Origins and Development*. Basel: Birkhäuser.

B. Russell. 1967–1968. *The Autobiography*, vols. I and II. London: Allen & Unwin.

B. Russell. 1971. *Russell: the Journal of the Russell Archives*. McMaster University.

B. Russell. 1983– . *Collected Papers*, about 30 vols, especially vols. 1–8. Routledge.

J. Vuillemin. 1968. *Leçons sur la première philosophie de Russell*. Paris: Colin.

Bertrand Russell (1872–1970), philosopher, mathematician, and political activist, was born in Trelleck, SE Wales. Russell studied at Cambridge where he became a Fellow of Trinity College in 1895. He published *Principles of Mathematics* (1903) and collaborated with A.N. Whitehead on *Principia mathematica* (1910–13). In 1916 his pacifism lost him his fellowship (restored in 1944), and in 1918 he served six months in prison. From the 1920s he lived by lecturing and journalism, and became increasingly controversial. Later works included *An Enquiry into Meaning and Truth* (1940) and *Human Knowledge* (1948). After 1949 he became a champion of nuclear disarmament, and engaged in unprecedented correspondence with several world leaders. One of the most important influences on twentieth-century analytic philosophy, he was awarded the Nobel Prize for Literature, and also the Order of Merit, in 1950.

Alfred North Whitehead (1861 – 1947), mathematician and philosopher, was born in Kent, SE England. Whitehead studied at Cambridge, where he was a Fellow in Mathematics at Trinity College until 1910. He then taught at London (1910–14), becoming Professor of Applied Mathematics at Imperial College (1914–24), and

then Professor of Philosophy at Harvard (1924–37). He collaborated with his former pupil, Bertrand Russell, in writing the three-volume *Principia mathematica* (1910–13). His philosophical works include *Adventures of Ideas* (1933) and *Modes of Thought* (1938). He received the Order of Merit in 1945.

Ivor Grattan-Guinness is holder of both the doctorate (Ph.D.) and higher doctorate (D.Sc.) in the history of science at the University of London, he is currently Professor of the History of Mathematics and Logic at Middlesex University, England. From 1986 to 1988 he was the President of the British Society for the History of Mathematics. Editor of the history of science journal *Annals of Science* from 1974 to 1981, in 1979 he also founded the journal *History and Philosophy of Logic*, editing it until 1992. He is an effective member of the *Académie Internationale d'Histoire des Sciences*. He edited a substantial *Companion Encyclopedia of the History and Philosophy of the Mathematical Sciences* (two volumes), 1994, London: Routledge. His latest individual books are *Convolutions in French Mathematics, 1800–1840*, three volumes (1990, Basel and Berlin: Birkhäuser); and *The Fontana* and *Norton History of the Mathematical Sciences. The Rainbow of Mathematics* (1997, London, Fontana; 1998, New York: Norton). In 2000 Princeton University Press published *The Search for Mathematical Roots, 1870–1940. Logics, Set Theories and the Foundations of Mathematics from Cantor through Russell to Gödel*. He is the Associate Editor for mathematicians and statisticians for the *New Dictionary of National Biography*, to be published in 2004.

11 Thomson, Rutherford and atomic physics at the Cavendish
BRIAN PIPPARD

One discovery after another in the last few years of the nineteenth century initiated so radical a change in the nature of physical science that it has become customary to distinguish between 'classical' and 'modern' physics. It was, of course, no overnight change – most active investigators continued along the lines that had brought them success, and encouraged their students to follow their example. A few, notably Max Planck, caught a glimpse of a new world and did not much like it. As usual, it was the unfledged young who broke with tradition, like Einstein who saw clearly what Planck had hardly appreciated. The process can be seen in a brief look at the first few Cavendish Professors of Experimental Physics, beginning with James Clerk Maxwell who had died well before the critical date of 1895 when Röntgen announced the discovery of X-rays. His successor Lord Rayleigh (who stayed only five years) was already in the middle of an immensely productive research career. His versatile mind continued to find new applications of the well-tried methods until his death in 1919 at the age of seventy-six, having published seven papers in his last year. The fourth Cavendish Professor, Ernest Rutherford, arrived in Cambridge from New Zealand as a new research student in 1895 at the start of the revolution and quickly developed into the greatest experimenter of his time and the pioneer of a new branch of science, atomic and nuclear physics. His most important work, however, was carried out at McGill University, where he became professor in 1898, and Manchester University to which he moved in 1907 before returning to Cambridge.

Between Rayleigh and Rutherford, an interval of thirty-five years, there was Joseph John Thomson who was elected to the Cavendish Chair in 1884 when Rayleigh retired to his family seat and private laboratory at Terling. Thomson's appointment at twenty-eight

– to his own surprise and the disapproval of some of his seniors, for his published record of research was still rather slender – shows what an impression he had made reading for the Mathematical Tripos and in his first years as a Fellow of Trinity. He was neither athletic nor well-off, and in a photograph with his student contemporaries looks a shy young man; yet according to the accounts of those more assured companions he was the one, among them all, they most confidently expected to achieve great things. He was unaffectedly friendly to everyone, he enjoyed jokes and gossip, and botanising on his walks in and around Cambridge; an ordinary, even a naïve youngster, one might think, were it not for the intellectual gifts and restless imagination that led him to become Cavendish Professor, Nobel laureate and Master of Trinity – but still JJ to all, even his family.

Like his predecessors in the chair he was by training a mathematician – that was the normal route to academic physics – but unlike Maxwell and Rayleigh he had on appointment little experience and less skill in practical matters. He was so notoriously clumsy that his students eagerly sought his advice so long as it stopped short of lending a hand. It seems almost from a sense of duty that he undertook his first serious experimental programme, but he did not cease to apply his mathematical skill to any question that caught his interest. A disciple of Maxwell, he developed his own approach to electromagnetic problems, and made an important advance when he showed that a moving sphere, carrying an electric charge, would generate a magnetic field by virtue of which its energy would be enhanced. It was as if some of the mass of the sphere resided in the surrounding ether, so that when the sphere moved the ether was the seat of some, at least, of the momentum and kinetic energy. This finding excited the notion that perhaps all mass resided in the ether, which might then be regarded as the sole origin of everything in the physical world (an idea shared with Kelvin who fancied ethereal vortices as models of material particles). At first there was no way for JJ to test this, but later developments gave him more confidence. Oliver Heaviside showed that the faster a charged body moved the more its mass would be enhanced, to the point that it

could not be accelerated beyond the speed of light. And when, some years after the discovery of the electron, Heaviside's expression was verified experimentally, it seemed to JJ that there was no place for an intrinsic mass apart from the electromagnetic component in the ether. He expressed this view explicitly in 1903, naturally unaware that Einstein's theory of relativity would soon demand the same enhancement of mass whatever its origin. To the end of his life JJ admired relativity theory as elegant mathematics, but held to the conviction that the all-pervasive ether was a necessity and Maxwell's theory a deeper statement of the truth. As he began, in the early years of the twentieth century, to devise atomic models his classical convictions encountered serious embarrassments which he recognised but managed to disregard, while his imagination thrust onward. In the ambivalent search for new insights while retaining utter loyalty to old revelations he stands Janus-like at the gateway to a new culture.

In the laboratory JJ chose to investigate the conduction of electricity in gases at low pressure. The electric discharge, with bands of glowing colours and dark striations, was spectacular but hard to reduce to such reliable measures as encourage mathematical analysis. Unskilled in glass-blowing and relying of necessity on primitive and slow hand-operated vacuum pumps, he needed an adept and patient assistant. Ebenezer Everett joined him from the chemistry department as a lad and stayed for forty-three years, sad not to outlast his professor but rewarded by the university with an honorary MA. There were no spectacular discoveries in the first ten years, but Röntgen's announcement of X-rays changed everything. The world was entranced by the sight of bones through living flesh, the physicians enlightened (and sometimes disconcerted) by a new diagnostic tool; the mathematically inclined physicists wrangled over the nature of the rays (transverse or longitudinal waves, or even pulses?); and the experimenters found with delight that X-rays made a gas conduct electricity. JJ was involved with everything, lecturing to the university with demonstrations barely two months after the announcement, X-raying patients brought in by local doctors, providing a sound theory of the origin of the rays,

publishing the first account of their ionising power, and revising his experimental programme to make the most of a great technical advance. Instead of applying a high voltage to start the discharge and ionise the gas it could now be done without any such fierce shock, and systematic measurements were possible. The way forward was clear, but strangely the discovery for which JJ is chiefly famous owed little to X-rays beyond stimulating his thought.

It had been known for decades that in a low-pressure gas discharge a radiation was emitted from the negative electrode – the cathode rays – which caused the glass of the tube to fluoresce. The rays propagated in straight lines like light, and cast shadows, but unlike light they were deflected by a magnetic field. They would have been taken for a stream of negatively charged particles but that they were apparently undeflected by an electric field. The general view in Germany was that they were waves, in England that they were indeed particles. JJ (or was it Everett?) managed to obtain a low enough pressure, as no one had done before, so that when an electric field was applied it was not neutralised by stray charges in the residual gas. Measurements in 1897 of the electric deflection of the cathode rays clinched the argument for negatively charged particles and showed that they were about a thousand times lighter than a hydrogen atom, the lightest known. At the same time Wiechert and Kaufmann in Germany had both independently reached the same conclusion, but JJ's special importance lay in his demonstration that the same particle was produced by electrodes of different metals, and in a variety of gases; moreover he and his students extended the work in ways that the others did not, and made it a special task to measure the properties of the new particle they called a corpuscle, clearly a basic constituent of all atoms. They concentrated on the corpuscle's electric charge and concluded that it was close to, and probably the same as, the charge carried by an ion in a salt solution; this charge had earlier been christened *electron* by Johnstone Stoney, and eventually the name was transferred to the corpuscle. The method used by JJ's students involved weighing and counting charged water drops; it was difficult and not

very accurate, but in 1909 Millikan thought of using oil instead of water and achieved impressive results in Chicago with drops that lasted for hours without evaporating. For a long time his value of the electronic charge figured in tables of atomic constants.

Once JJ was assured that all atoms contained negative electrons he began to devise detailed models and quickly ran into difficulties. Since atoms are electrically neutral there must be positive charges as well as electrons, but they never appear in anything like the same form as electrons; the positive particles that are observed carry all the mass of an atom and are indivisible. If they are single particles and their mass is electromagnetic they must be even smaller than electrons, for the electromagnetic mass is inversely proportional to the radius. In such a case the atom would be a collection of tiny particles held together by electrostatic attraction. But it had long been known that such a collection cannot be stable if all parts are at rest, while if they are in motion they will radiate energy and fall together. Although JJ's lectures, at Princeton University in 1903 and at the Royal Institution in 1906, are a little vague about details (characteristic of one who attached less weight to ascertained truth than to the stimulating power of bold ideas) he seems strongly attracted to the model Kelvin had recently published under the title *Aepinus atomized* and which became known as the currant-bun model. The atom is considered as a sphere of uniform charge density in which the electrons are free to move; in such an environment they can find stability at rest or in motion. If at rest they may occupy nesting shells whose pattern tends to repeat periodically as more electrons are added – behaviour suggesting an explanation of Mendeleev's Periodic Table of the elements. If the electrons move, they will radiate energy, but perhaps only slowly; here is a possible mechanism of radioactive decay but this suggestion was an example of JJ's optimism, as was his disregard of awkward details – what holds the positive sphere together, and if it is as large as an atom where does its mass come from? There was enough to do without tackling such matters, which soon became irrelevant with Rutherford's invention of the nuclear model; but if these puzzles had no other effect

at least they concentrated JJ's attention on the question of how many electrons there are in an atom.

His original thought was that all the mass was contributed by the electrons, thousands in even the lightest atoms. This complexity had its attraction for any one attempting by classical arguments to explain the great number of spectral lines emitted by even the simplest atoms; but JJ himself dealt the death blow to the idea. His analysis of three different types of experiment persuaded him that an atom could not have more electrons than its atomic weight. One of the experimental results, on the scattering of X-rays, was that of Charles Barkla, a talented physicist (Nobel laureate 1917) and so excellent a bass in King's College choir that he was offered a substantial scholarship in the vain hope of preventing him going to Liverpool University. The effect he investigated and JJ analysed is still known as Thomson Scattering, and is basic to crystal analysis by X-ray diffraction. Further scattering experiments reduced the likely number of electrons in an atom to the point of rough agreement with an independent suggestion that it was the same as the atomic number. After the work of Rutherford and his team in Manchester, 1911–14, and Bohr's atomic theory of 1913 there was no room for doubt, and if JJ felt his classical difficulties had been swept under the carpet (as they had) he gave no sign, for other exciting questions had captured his attention.

Decades before in Berlin, Goldstein had found that holes cut in the cathode of a discharge tube let through positively charged particles, which JJ identified in 1905 as ionised atoms, much more massive than electrons. Deflecting them with parallel electric and magnetic fields he obtained photographic records in the form of parabolic traces, each one corresponding to a different mass. His new colleague Francis Aston was a skilful experimenter who made immediate technical improvements to the quality of JJ's work. By 1914 they had shown the presence of many unsuspected chemical species in an ionised gas, as well as two ions of mass 20 and 22 when the gas in the tube was neon. Rutherford and Soddy at McGill, in their work on radioactivity, had discovered the existence of isotopes, chemically identical but differing in mass. There

were no examples known, however, of isotopes of lighter elements, and the two parabolas for neon provoked disagreements that persisted until after the First World War; Aston was sure they were separate isotopes, JJ and others favoured a single form of neon and a compound, say neon hydride. When Aston returned in 1918 and made his first mass spectrograph he was able to establish that not only neon, but almost all the many elements he examined, occurred in several isotopic forms. He devoted the rest of his working life to improving the design of his instrument and measuring isotopic masses with high precision. This paid handsomely when, after 1932, nuclear transmutations became a central feature of experimental physics and Einstein's $E = mc^2$ related a small decrease in mass to a large production of energy.

We have now passed the date, 1918, when JJ relinquished the Cavendish chair to become Master of Trinity. His great period of productive research had ended at the outbreak of war and, although he still visited the laboratory regularly, and never ceased his attempts to find classical interpretations of new results, he was no longer an influential figure in physics. In the last years before his death in 1940 the newer researchers hardly recognised in the old shuffling figure the man of whom Aston wrote 'His boundless, indeed childlike, enthusiasm was contagious and occasionally embarrassing'; this vivacity, together with his inventiveness and an extraordinary ability to detect and overcome flaws in the experiments of others, played a major part in the evolution of a laboratory that became in his day a byword for excellence.

For all its fame the Cavendish cannot have seemed to Rutherford an easy place to take over in 1919. He had been marvellously productive at McGill and Manchester, but the Cavendish had nearly come to a halt during the war and in addition to chronic penury was seriously short of talent. JJ thought he could combine the direction of research with the Trinity mastership and had to be tactfully dissuaded. It was perhaps only Rutherford's respect for his old professor that swayed him, at the last possible moment, so that the electors to the chair could appoint the one candidate for whom they felt any enthusiasm; and rightly, for though he had little opportunity to continue making

personal contributions on the same scale as before, his driving force gave new and powerful stimulus to the research. In his absence since 1898 he had done more than anyone to create nuclear physics. Before he left he had observed two types of particles emitted in radioactive decay and had christened those most easily stopped by thin foils, the α-rays (which he loved above all) and those that went further, the β-rays. Within the next few years he had identified α-rays as ionised helium atoms and β-rays as electrons (a third type of radiation, the γ-rays, were found later and eventually identified as X-rays of very short wavelength). Rutherford and Soddy sorted out the diverse chains of radioactive processes which converted uranium to lead; from the chance observations of Marsden and the careful measurements that followed, he had developed his nuclear model; and during the First World War had observed and interpreted the disintegration of the nitrogen nucleus by a fast α-particle, with the emission of a proton.

All this, and much more, he had done since Becquerel's discovery of radioactivity in 1896, a year after he reached Cambridge from New Zealand. He could hardly have chosen a better moment to arrive, especially as he brought with him the radio detector he had invented; by demonstrating communication over a mile or more he won immediate notoriety and a following among the academic élite. He showed confidence and authority, combined with an easy affability, for all that his contemporary, Paul Langevin, when asked later how they had got on, said, 'One can hardly speak of being friendly with a force of nature'. A few months after his appearance in the Cavendish, X-rays were announced, and JJ soon deflected him into studying the ionisation of gases, particularly the movement and recombination of ions. This first venture shows he would have done great things in the physics of ions, but radioactivity was too strong a temptation and it was his work on the radioactive series that won him the Nobel Prize (for chemistry) in 1908. It was a mature Rutherford who returned in 1919, sensitive to the needs of others behind a manner that was direct to the point of brusqueness, with rare eruptions into terrifying anger for which he was immediately and sincerely apologetic; avid for results, and in his impatience

apt to haul photographs prematurely from the fixing bath for a first look at something new (JJ also had this weakness); always ready to applaud success and encourage quick publication.

Rutherford's arrival in 1919 was at a moment of fine opportunity, for there were men of great gifts looking for the chance that the war had denied them, and the prospect of joining Rutherford was just what they needed. James Chadwick, a promising research student at Manchester, had been awarded a scholarship in 1913 on condition he moved elsewhere, and Rutherford had arranged a place in Berlin with his old associate Hans Geiger. When war came he was interned with other stranded enemy aliens and did not return home till 1918, starved but eager to resume the life of science. He came to Cambridge with Rutherford as a senior research student and rose to serve as his deputy until 1935 when he was appointed to the chair at Liverpool. His last years in the Cavendish had been outstandingly successful, as will be told later, but he wanted to run his own show and disagreed with Rutherford's reluctance to build a cyclotron – recently invented by Lawrence in California but already recognised as a most promising tool for nuclear research. The Liverpool cyclotron was eventually finished (by others and after the Second World War) when Chadwick's research career had come to an end with his involvement in nuclear politics and his influential representation of British interests in the American atomic bomb development – an extraordinary success for a man apparently handicapped by reticence and self-doubt.

During Chadwick's internment from 1914–18 he was befriended by Geiger and other German scientists to the extent that he could carry out simple experiments. These caught the interest of a younger internee, Charles Ellis, who had been enjoying a holiday in Germany after his first year at the Sandhurst Military Academy. On release he entered Trinity Colleage, graduated in 1920 and started research at the Cavendish where he remained until 1936, making notable contributions towards sorting out the problem of β-decay. The war had a similar effect on Patrick Blackett who served as a naval officer, and would have continued but for the chance of being seconded to Cambridge to repair

the damage his education had suffered; he began his varied and distinguished research career in 1921. Much of the work that earned him a Nobel prize was carried out before he left the Cavendish in 1933, but his later pioneering research in geophysics may be considered his most significant scientific contribution.

Chadwick, Ellis and Blackett are three notable examples of the new intellectual vigour brought to the Cavendish by Rutherford after 1919. The only contemporary of Rutherford's research student days still in the Cavendish was C.T.R. Wilson; in Blackett's words 'of the great scientists of the age, he was perhaps the most gentle and serene, and the most indifferent to prestige and honour'. Entranced as a youth by the sunlight and mists on Ben Nevis, he devoted himself to clouds and thunderstorms, and with skill and imperturbable patience sorted out the processes by which water vapour condenses. Moist air blowing up the mountainside is cooled by expansion and becomes supersaturated, but does not readily condense except on to dust particles. In his laboratory experiments Wilson cleaned the air of particles by repeated expansions, and was able to verify and make quantitative JJ's earlier prediction that electric charges would encourage condensation. It was from his meteorological studies that he devised the cloud chamber for making visible the tracks of charged particles ejected in radioactive decay. The drops that condensed along the trail of ions left by the particle rapidly grew and fell, and had to be photographed immediately after the expansion. An instrument like this, that removed the interpretation of events from the realm of speculation, appealed strongly to Rutherford, and as a new research student Blackett's task was to use one to obtain direct visual evidence of the disintegration of a nitrogen nucleus by an α-particle. Many photographs were needed before one of these rare events appeared, and much thought went into improving the efficiency. A breakthrough was achieved some years later by Blackett and Giuseppe Occhialini who 'had come for three weeks and stayed for three years', bringing from Florence the new technique of coincidence counting. By placing Geiger counters above and below the cloud chamber, expanding and photographing only when fast particle triggered both, they

observed without wastage relatively rare events like the passage of energetic cosmic ray particles. As well as making advances in cosmic ray studies, they confirmed irrefutably Carl Anderson's single observation of a positron produced by cosmic rays. The positron, with properties like an electron but positively charged, had been reluctantly predicted by Paul Dirac as a necessary consequence of his relativistic quantum mechanics; he later felt that Blackett had been unnecessarily cautious in delaying publication of his photographs until he was quite sure of their interpretation, and thus ceding priority to Anderson.

The phenomenon photographed many times by Blackett and Occhialini was the production of a positron-electron pair when a sufficiently energetic particle collides with a nucleus. If this had only forestalled Anderson, it would have been one of a trio of important discoveries by Cavendish researchers in 1932. The other two were the first disintegration of a nucleus by artificially accelerated particles and Chadwick's discovery of the neutron. Both were indebted to the commanding inspiration of Rutherford, who saw in his nitrogen disintegration a means of probing nuclear structure but could make little progress with α-particles as missiles; doubly charged, they were too strongly repelled by their targets. In 1927 John Cockcroft had just taken his Ph.D. and, when he heard George Gamow explain his theory of α-particle emission by quantum-mechanical tunnelling out of a nucleus, he saw that the reverse process might allow protons of moderate energy to penetrate the barrier. He had no difficulty persuading Rutherford to get a little money for the venture. Ernest Walton who had his own ideas for accelerating particles, but was finding them difficult to realise, joined forces with Cockcroft to put together a generator for accelerating protons in a vacuum through a potential drop of half-a-million volts. The picture shows the typically (for the Cavendish) rough-and-ready construction with Walton crouched in the lead-covered tea chest below the evacuated tube. If the protons striking the target at the bottom cause nuclear fragments to be thrown off, the screen he is observing will scintillate where the fragments hit. One day in April 1932 Walton was working alone (it would not be allowed nowadays)

FIGURE 11.1 E.T.S. Walton in the lead-covered tea-chest below the evacuated tube through which ions are accelerated. In the can at the top of the tube is the ion source at about 500,000 Volts positive with respect to the observation chamber. The ions (protons) are produced by a discharge through hydrogen at a low pressure. The figure on the left is not Cockcroft. *Source:* Cavendish Laboratory Archives.

because Cockcroft was helping Kapitza with a quite different project. He put in a lithium target, switched on the high voltage at the other end of the room, and returned to the box crawling to avoid being at the receiving end of a spark. The screen was scintillating; Cockcroft came as soon as he could, and Rutherford came the next day to see for himself and give his blessing to Walton's view that a proton entering a lithium nucleus split it into two α-particles. Aston's mass-spectrograph gave accurate enough masses of all the nuclei concerned to show that a little mass disappeared in the reaction, and when they measured the energy of the α-particles they found it agreed well with Einstein's $E = mc^2$; over twenty-five years since it was derived, this basic expression received its first explicit verification. Its use in nuclear physics is now commonplace and we shall meet another example soon. Of more obvious significance is the world-wide development of ever bigger accelerators – cyclotrons, synchrotrons, linear accelerators – which were foreshadowed by the modest, home-made contraption of Cockcroft and Walton.

After two years of intensive work, they left academic research and followed very different careers. Walton went back to his old university, Trinity College, Dublin, where eventually he became Professor of Physics and was greatly admired for his lectures and demonstrations to students; Cockcroft became a distinguished scientific administrator and eventually the first Master of Churchill College. Both were notably unostentatious – in Walton it was true modesty, in Cockcroft something more like calm confidence; his smile seemed at times the mask of an iron will, and he possessed a formidable power of silence.

Very shortly after the first disintegration came Chadwick's paper on the neutron. Ever since the nuclear model of 1911 there had been speculations about its detailed constitution, with protons and electrons as the components, no other particle being known. In 1920 Rutherford suggested that a proton and an electron might be closely bound in a neutral pair, but the idea was not sanctioned by Bohr's theory of the hydrogen atom and there was no supporting evidence, so it remained in limbo. In 1930 fast α-particles striking beryllium were found to generate a highly penetrative radiation which was taken to be energetic

γ-rays. When Irène Curie and Frédéric Joliot showed that this radiation could eject fast protons from a target of paraffin wax, it had to be allowed that the energy of the γ-ray quantum was very high indeed. Chadwick found the whole notion untenable, especially when all the light elements he tried could be knocked on as readily as protons. He knew, of course, Rutherford's dream of a neutral particle, and gave simple and clear arguments why this was a much better explanation than γ-rays. Moreover, on this assumption he could estimate the masses and energies involved, and conclude that the new particle, the neutron, was a little lighter than a proton and an electron together, consistent with its being a rather weakly bound pair. The proposal was readily accepted and it was soon clear that neutrons could be made in several nuclear reactions. The Italian school, led by Enrico Fermi, began bombarding different nuclei with neutrons (which, being uncharged, entered a nucleus without hindrance) and produced many transmutations. It was this work that led, a few years later, to fission of the uranium nucleus and all that followed – but this is outside our present theme.

A young American theorist, Maurice Goldhaber, pointed out to Chadwick that the recently discovered deuteron, presumably made up of a proton and a neutron, was so weakly bonded that it could be split by available γ-rays. Chadwick invited him to share the test of his idea, and from the energy of the resulting proton and neutron make a better determination of the neutron mass ($E = mc^2$ again). Their result was a little larger than Chadwick's original estimate, more even than the combined mass of a proton and an electron so that Goldhaber, by his own account, was shocked to realise that a neutron might decay into a proton and an electron, together with a neutrino which Pauli had recently proposed as a desirable concomitant of radioactive β-decay (and which Ellis and Nevill Mott kicked themselves for not having thought of first). Goldhaber had enough information to estimate that the lifetime of a free neutron should be half-an-hour or less, but only after the Second World War was there sufficient flux of neutrons, from a nuclear reactor, to make a direct determination possible; the lifetime is now known to be a little over ten minutes.

This account has concentrated on the work nearest to the hearts of Thomson and Rutherford, but much more was happening in the Cavendish, and not all in atomic and nuclear physics. Rutherford has acquired a reputation for intolerance of other branches of science, but he was happy to find space in the laboratory for congenial men of high intellect; Peter Kapitza, G.I. Taylor, Edward Appleton, and others were welcome, and sowed the seeds for a post-war flowering. By 1945 America was poised to take over world leadership in the physics of nuclei and fundamental particles, and for many years no other country had the organisation or the money to compete. The Cavendish has continued to work in particle physics and has achieved much of value, but the great days ended with Rutherford's death and the outbreak of war.

Lawrence Bragg succeeded Rutherford and, like him, had to rebuild a research team seriously depleted by the six years of war. He was not a nuclear physicist but had invented X-ray crystallography at the age of twenty-two while a research student at the Cavendish. His return gave new impetus to crystallographic research in Cambridge, and his encouragement was an important factor in the prolonged efforts of Max Perutz and John Kendrew to determine the arrangement of atoms in protein molecules. Of Rutherford's non-nuclear colleagues Kapitza was no longer in Cambridge, having been detained in Moscow when he made one of his regular visits in 1934. The helium liquefier he had designed became the central facility for the work of Jack Allen and David Shoenberg who established low temperature research in Cambridge with important discoveries in superfluid helium and superconductivity. It was Shoenberg's devoted studies of the motion of electrons in metals, in which I also was involved, that provided a sound experimental basis for what has become a highly developed quantum theory of electrons in crystalline lattices. G.I. Taylor remained in Cambridge till his death in 1975, an exceptionally versatile and imaginative classical physicist, one of the discoverers of the dislocations which limit the strength of materials, a pioneer in understanding that most difficult phenomenon, turbulent flow of liquids, and the inventor of the CQR anchor which is now standard equipment in small craft.

The third member of the group, Appleton, spent two periods in Cambridge; he started the Cavendish work on radio propagation after the First World War and Jack Ratcliffe was one of his students. Appleton left, for the second time, in 1938 and Ratcliffe returned in 1945 to build a strong team in ionospheric research, with Martin Ryle and Graham Smith among them. From this beginning sprang radio-astronomy which has enormously increased our knowledge of stars and galaxies, and of the cosmos almost from its primal explosion. These are some of the achievements of Bragg's and his successor Mott's tenure of the Cavendish chair, justifying the claim that the post-war Cavendish has been no less significant as a source of new ideas than it was in the now-legendary days of Thomson and Rutherford.

Further reading

G.K. Batchelor. 1996. *The Life and Legacy of G.I. Taylor*. Cambridge: Cambridge University Press.

A. Brown. 1997. *The Neutron and the Bomb: A Biography of Sir James Chadwick*. Oxford: Oxford University Press.

Lord Rayleigh. 1942. *The Life of Sir J.J. Thomson*. Cambridge: Cambridge University Press.

J.D. Watson. 1999. *The Double Helix*. London: Penguin Books.

D. Wilson. 1983. *Rutherford: Simple Genius*. Boston, MA: MIT Press.

Sir Joseph (John) Thomson known to all as JJ, was born (1856) in Manchester where at the age of fourteen he entered Owens College, now Manchester University. A scholarship enabled him in 1876 to begin reading mathematics at Cambridge where he graduated as Second Wrangler and was soon elected to a Fellowship at Trinity College. His early researches, on vortex rings but especially on electromagnetic theory, won him the Cavendish Chair of Physics when Lord Rayleigh resigned in 1884. He was only twenty-eight, and occupied the chair for thirty-five years, gathering colleagues and research students of such quality, and inspiring them with such ideas and enthusiasm as to bring worldwide fame on the Cavendish Laboratory. Much of the laboratory's work concerned conduction of electricity in gases – the nature of ions, their production, recombination and mobility. His personal fame, however, rests more on related investigations, alone or with students: the discovery of the electron (1897), the earliest measurements of its charge, the demonstration of how many fewer there are in an atom than he had first

thought, and the positive ray parabolas which, in Aston's hands particularly, led to the mass spectrometer and the cataloguing of an unsuspected plenitude of isotopes throughout the Periodic Table. In 1919 JJ was elected Master of Trinity College and resigned the Cavendish Chair to Rutherford. He made no more advances in physics after that, but devoted himself to his College until his death in 1940.

Ernest Rutherford, created Baron Rutherford of Nelson in 1931, was born (1871) near Nelson on the South Island of New Zealand. After graduating at Canterbury College he won (1895) an 1851 Exhibition to Cambridge, entered Trinity Colllege and became the first Cavendish research student to have graduated elsewhere. He was already a pioneer of radio transmission, but was persuaded by J.J. Thomson to use the newly discovered X-rays to produce ions in gases and study their mobility. Both these research fields offered scope for his talent, but radioactivity was even more inviting, and before moving to Montreal as Professor at McGill University (1898) he had already distinguished and named the α and β emissions. He and Frederick Soddy sorted out the chains of distintegration by which uranium decays into lead. His move to Manchester in 1907 did nothing to diminish his activity; with colleagues and students he showed definitely that α-particles were doubly ionised helium, and proposed his nuclear model of the atom to account for their scattering through large angles. His last research in Manchester, before taking up the Cavendish Chair in 1919, was the distintegration of nitrogen nuclei by α-particles, a discovery that marks the beginning of nuclear physics. His first appointments to the physics staff in Cambridge more than made good the loss of talent during the First World War. Until his death in 1937 he presided over a grand succession of advances in nuclear physics, too much involved in outside affairs to play an active role himself, but still and always the world leader, respected, loved and (a little) feared.

Sir (Alfred) Brian Pippard, was born (1920) in London. He took a physics degree at Cambridge and, except for four wartime years working on radar and one sabbatical year in Chicago, has spent his whole adult life there. His research has been largely on electrical conduction in metals at low temperatures, notably high-frequency measurements on superconductors which led to a non-local extension of the London equations, and determination by the anomalous skin effect of the Fermi surface of copper. He succeeded Sir Nevill Mott as Cavendish Professor in 1971, and was the first President of the new Graduate College, Clare Hall, founded in 1966.

12 Hopkins and biochemistry
HARMKE KAMMINGA

Frederick Gowland Hopkins has gone down in history as 'the father of British biochemistry', and it was largely through his efforts that the Cambridge Department of Biochemistry became a centre of world renown. The department was established in 1914 and Hopkins was its first Professor, until his retirement at the age of eighty-two in 1943. Between the two world wars, Hopkins and his colleagues put into place an exceptionally wide-ranging programme of biochemical research, developed degree course teaching and research training in the subject, and hosted visiting researchers from every continent. By the time of Hopkins' death in 1947, some seventy-five former members of the department had been elected to professorial chairs worldwide.

While Hopkins' legacy is celebrated among biochemists, his career in Cambridge did not run a uniformly smooth path. Before he acquired his own department, Hopkins had to struggle very hard to find the time, laboratory space and resources to do the biochemical research he wanted to pursue in the Cambridge Physiological Laboratory. Later, in the 1920s, concerted efforts were made to place the leadership of his department in other hands, albeit unsuccessfully. In both cases, Hopkins' particular ambitions for biochemistry went far beyond the expectations, or indeed wishes, of his peers. Here I shall outline these ambitions, the contexts within which Hopkins pursued and defended them, and some of the obstacles that faced him.

FROM PHYSIOLOGY TO BIOCHEMISTRY

Hopkins was trained in analytical chemistry and subsequently qualified at medical school. He began his research career in the 1890s as a demonstrator in physiology at Guy's Hospital, London, where he built up a strong reputation for his skilful chemical analyses of substances of

physiological and pathological importance. Here he also became interested in the serial chemical transformations undergone by biological substances in the course of metabolism.

In 1898, the renowned Cambridge Professor of Physiology, Michael Foster, invited Hopkins to join his Physiological Laboratory as a lecturer in chemical physiology. The University did not allocate extra funding for this post, however, and Hopkins had to supplement the meagre stipend he received by taking on a very heavy college teaching load. Besides coping with these pressures on his time, Hopkins had to compete continuously with his colleague physiologists for research funds and facilities. This problem became increasingly acute after Foster's retirement in 1903. Foster had long propagated a broad and inclusive view of physiology, as part of his – successful – attempt to establish physiology as an autonomous science in the University. Nevertheless, research in his own department was focussed overwhelmingly on physical, as opposed to chemical, approaches to physiological problems.

For Foster's younger colleagues, these physical approaches had become the norm and his successor, J.N. Langley, seems to have regarded the chemical investigations Hopkins pursued largely as a drain on valuable resources. A separate chair for Hopkins soon became a desirable goal for all concerned, the first moves in this direction being made as early as 1904. Langley and other senior physiologists played an active role in the protracted negotiations at different levels of the University's complex bureaucracy. Despite very strong support for the proposal of a Department of Biochemistry, Hopkins' hopes were dashed repeatedly, but in 1914 the goal was finally reached.

Although Hopkins felt increasingly alienated in the Physiological Laboratory, this was the site where he carried out most of the research on which his eminent scientific reputation came to rest, especially in the biochemistry of nutrition. This work was rooted in investigations of the amino acid composition of proteins that Hopkins had carried over from London. In 1901, Hopkins and Sydney Cole isolated in pure form the hitherto rather obscure substance tryptophan and identified it as an

FIGURE 12.1 Hopkins (right) holding the first pure sample of tryptophan, 1901; with Sydney Cole (left).
Source: From the collection of the Department of Biochemistry, Cambridge.

amino acid. In animal feeding experiments done together with Edith Willcock, Hopkins found that tryptophan must be supplied in dietary protein for rats to grow normally: when the protein component of their diet lacks tryptophan, they stop growing and eventually die, while normal growth resumes upon addition of tryptophan to these diets.

This observation undermined the prevailing consensus that all proteins are nutritionally equivalent. It had been assumed that the main nutritional function of proteins was to supply nitrogen, which is a component of every amino acid. The tryptophan experiments suggested that, on the contrary, different amino acids have different functions in animal metabolism. This motivated a new line of research for Hopkins, in which he investigated the dietary requirements of animals for individual amino acids, leading to the identification of several more 'essential amino acids'.

As a sideline to this work, Hopkins formulated the notion of 'accessory food factors', on the basis of systematic feeding experiments using diets consisting of pure components of known composition. In a public lecture in 1906, he concluded that normal animal growth requires minute quantities of as yet unknown 'accessory food substances', in addition to pure carbohydrates, fats, proteins, minerals, and water. In the same lecture, he speculated that diets deficient in these hypothetical substances might be implicated in diseases such as rickets and scurvy, for which a dietary link had long been suspected.

In 1912 Hopkins published the full results of a large number of carefully controlled feeding experiments in support of the existence of accessory food factors, in a long paper in the *Journal of Physiology*. For Hopkins, the most significant feature of this work was that it demonstrated clearly that the quantitative energy (or calorie) content of diet is not a sufficient indicator of its qualitative adequacy. This made it imperative to move beyond crude input/output studies and to investigate the role of individual, chemically clearly defined nutrients in intermediary metabolism. Thus Hopkins stressed the importance of biochemical approaches.

Hopkins' accessory food factors became more widely known as

'vitamins'. A vast and diverse research effort in different countries was devoted to the elucidation of their chemical properties and physiological role, as well as their therapeutic use in the new disease category of vitamin deficiencies. The first rigorously demonstrated case of a specific dietary deficiency disease was beri-beri, based on studies in the 1890s by the Dutch hygienist Christiaan Eijkman that were subsequently reinterpreted in the language of vitamins. In 1929, Hopkins and Eijkman shared the Nobel Prize for Physiology or Medicine, for the 'discovery of vitamins'.

This was Hopkins' most widely celebrated scientific achievement. He was already seen publicly as an authority on the biochemistry of nutrition before a separate Department of Biochemistry was established. His work on accessory food factors had received wide publicity in the national press, and he was appointed as co-director of a new, government-funded Institute of Animal Nutrition in Cambridge in 1912. During the First World War, he served on committees concerned with war-related nutrition research and national food policy, and in 1918 he was appointed Chairman of the Accessory Food Factors Committee of the Medical Research Committee (subsequently Medical Research Council). This committee issued a number of authoritative reports on the state of knowledge concerning accessory food factors or vitamins, the first of which appeared in 1919. In the historical introduction, Hopkins was presented as the discoverer of these substances, his 1912 paper being described as a classic contribution. In many of his public lectures of the 1920s and 1930s, Hopkins devoted attention to vitamins as an illustration of the medical and social importance of biochemical research.

Hopkins was widely expected to take a leading part in elucidating the precise metabolic role of vitamins. Once he had the powers to steer the direction of research in his own department, however, Hopkins pursued much grander ambitions for biochemistry. He had first articulated these ambitions in public in 1913, in a famous address to the British Association for the Advancement of Science.

In this lecture, entitled 'The dynamic side of biochemistry',

Hopkins presented an inspiring vision of biochemistry as a young science concerned with the chemical study of all biological processes, a science that combines chemical rigour with 'biological instinct'. For Hopkins, biochemistry should be neither simply a handmaiden of medicine, nor concerned solely with the chemical identification of molecules isolated from animals. Above all, it should be centred on the chemical transformations undergone by these molecules in the basic processes of life, using a combination of organic chemical analysis and the physicochemical study of equilibrium dynamics. Presenting a view of the living cell as an organised, highly differentiated system of interdependent chemical processes in dynamic equilibrium, Hopkins set out a programme for unravelling the chemical reactions of the cell, their spatial organisation, and the means whereby they are controlled and coordinated. He stressed that this seemingly forbidding chemical complexity was open to investigation: biochemistry deals with simple molecules undergoing comprehensible reactions. He urged young chemists to take up biochemical research, emphasising how much their technical skills were needed in this 'borderland' between biology and chemistry.

Still regarded as a seminal contribution, Hopkins' programmatic lecture was widely publicised at the time. In 1913, however, he was not yet in a position to put into practice the ambitious programme of biochemistry that he envisaged, and soon after Hopkins was placed in charge of the new Department of Biochemistry in 1914, most of the department's work became devoted to the war effort. After the war, Hopkins' staff expanded greatly and research in the department soon began to follow the kind of lines he had advocated in 1913. In particular, he reoriented research away from 'applied' problems, such as those concerning nutrition, towards what he considered 'fundamental' investigations of biological catalysis, oxidation and reduction.

THE DEPARTMENTAL CULTURE AND ITS CRITICS
From the 1920s, research in Hopkins' department covered an enormous range of topics, including the control of specific metabolic reactions by

a multitude of enzymes, formal enzyme kinetics, tissue respiration, muscle biochemistry, bacterial metabolism, plant biochemistry, biochemical embryology, and comparative biochemistry. To many outsiders, the sheer diversity of this research seemed bewildering. For Hopkins, however, there was an underlying unity to the programme as a whole, and he repeatedly presented the work done in his department as exemplifying his perspective of biochemistry as a fundamental, unifying science of life centred on cellular dynamics. His colleagues later often stressed his remarkable skills in articulating the relationships between the narrowest individual lines of research and larger questions of biological function. Through his participation in the weekly research talks in the department, he is said to have given every investigator the sense that they were not pursuing small, isolated problems, but contributing to a coherent programme of research.

Hopkins' department flourished not only at the level of research, but also in terms of interpersonal relations. In reminiscences written by those who had direct experience of working in the department, Hopkins' group is invariably presented as a happy family, presided over by a benign paterfamilias. Hopkins gave his colleagues a great deal of freedom in choosing their own research topics, many of them engaged in fruitful collaborative research, and they kept abreast of each other's work. Several of them disseminated the work in their own research fields in highly regarded books, several of which were the first synthetic works in their particular subject area. They also published a humorous in-house journal called *Brighter Biochemistry*, to which Hopkins, too, contributed.

Many of Hopkins' colleagues acquired great scientific distinction in their own right. For instance, J.B.S. Haldane was a Reader in the department for ten years before his appointment as Professor of Genetics at University College, London. While Haldane is best remembered for his work in evolutionary biology and genetics, in Cambridge he made important contributions to enzyme kinetics, a field further developed in major ways by Malcolm Dixon. Haldane also did pioneering work in biochemical genetics in Cambridge, developing

Muriel Wheldale's early work on the inheritance of plant pigments. Marjory Stephenson was internationally recognised as one of the chief founders of the field of microbial biochemistry, and was one of the first two women to be elected Fellows of the Royal Society, in 1945, together with London X-ray crystallographer Kathleen Lonsdale. Dorothy Needham became an international authority on the biochemistry of muscle, and Joseph Needham firmly placed biochemical embryology and morphology on the map. Ernest Baldwin wrote the first book ever on comparative biochemistry and, in 1947, published the very widely used textbook *Dynamic Aspects of Biochemistry*, which spread the dynamic approach of Hopkins' school to students worldwide. Baldwin dedicated the book to 'Hoppy'.

The full roster of people working in Hopkins' department includes too many eminent names to mention individually. One striking feature, however, is the unusually high number of women who contributed to the research and teaching of the department, those named above representing only a small proportion. Their number would have been remarkable in any university at the time, especially in the sciences, but the University of Cambridge was among the least encouraging academic environments for women. The University did not even admit women to the title of degree until 1921, and only allowed them to graduate formally from 1948. Yet in the Department of Biochemistry women were treated on an equal footing as scientists and colleagues from the start. Good collegial relations within the department even took the form of marriages between its members, notably between Dorothy Moyle and Joseph Needham, Muriel Wheldale and Huia Onslow, and Hopkins' daughter Barbara (who worked on brain metabolism) and Eric Holmes.

Also remarkable is the international composition of the community working in Hopkins' department at any given time. Overseas visitors joined the department from the start. In the 1930s, Hopkins was particularly active in the Academic Assistance Council, securing places and funds for refugee biochemists from central Europe, often in his own department.

All reminiscences written by biochemists who worked in Hopkins' department for any length of time stress its friendly and intellectually stimulating atmosphere, paying tribute especially to Hopkins' breadth of biochemical vision and the inspiration and subtle guidance he provided. While Hopkins' colleagues clearly regarded the 1920s and 1930s as the golden age of Cambridge biochemistry, this was the very time that Hopkins and his school came in for protracted criticism.

Neither the research priorities in Hopkins' department nor the freedoms accorded its members found uniform approval. Behind the scenes, powerful members of the medical-scientific establishment in and beyond Cambridge began to question Hopkins' entire style of leadership. The first to express his dismay at the turn biochemical research in Cambridge had taken, in private correspondence with Hopkins, was Walter Morley Fletcher, Secretary of the new Medical Research Council. This was a serious matter for Hopkins, as the two men were close allies who had long enjoyed a relationship of mutual support and admiration.

Fletcher and Hopkins had collaborated in the Physiological Laboratory, in celebrated investigations of metabolic events associated with muscle contraction, fatigue and recovery. When the Medical Research Committee was established in 1913, Fletcher was appointed its Secretary on Hopkins' recommendation. In 1919, the Committee was transformed into the central government funding body for medical research, the Medical Research Council (MRC), and Fletcher's political power and influence became instrumental in creating the climate and resources that allowed Hopkins' department to flourish. Not only did the MRC provide the department with research funds, it was Fletcher who successfully persuaded the Dunn Trustees to grant Hopkins a very generous bequest to build, staff, and equip a large institute, which was opened officially in May 1924 under the name of Dunn Institute of Biochemistry. This boost to resources and facilities heralded the end of the cramped laboratory conditions that the department had suffered until then, and enhanced Hopkins' institutional autonomy within the

University. He became even more determined to build up his school as a centre for fundamental research in 'dynamic biochemistry'.

Fletcher, on his part, adopted an increasingly proprietorial attitude towards Hopkins' affairs after he had secured the Dunn bequest in the early 1920s, repeatedly expressing his hopes for a large new research effort from Hopkins on vitamins. Hopkins himself had already turned away completely from vitamin research, in favour of investigations of the role of a molecule that he had isolated in 1921, glutathione, in biological oxidation and reduction processes. By the mid 1920s, vitamins had virtually disappeared from the entire departmental research repertoire. Fletcher was not opposed to basic research, believing that sound clinical practice required solid scientific foundations, but he did expect long-term medical objectives to be held in view. The medical importance of vitamins had already been demonstrated in the form of their therapeutic efficacy in dietary deficiency diseases such as scurvy and rickets, but their precise functions in intermediary metabolism remained obscure. To Fletcher, this seemed the perfect, and most pressing, biochemical research area for Hopkins and his colleagues.

Fletcher bemoaned the fact that Hopkins had admitted so many people to his department and allowed them to pursue so many different lines of research, the significance and coherence of which he questioned. Running such a large and diverse department, as well as serving on many committees, also meant that Hopkins spent far too much time on administration and far too little on research, the latter being his greatest strength in Fletcher's eyes.

When Fletcher's repeated attempts to persuade Hopkins to redirect his priorities proved unsuccessful, he took his concerns to a wider circle. In fact, his misgivings about the way in which Hopkins ran his department were shared by an informal network of influential scientists inside and outside the University of Cambridge. Nearly all were friends of Hopkins who had worked in the Cambridge Physiological Laboratory at some time, and included William Bate Hardy, A.V. Hill, Henry Dale, and Thomas Elliott. It also included Hugh Anderson,

Master of Caius College, who was an immensely powerful figure in university politics at the time. This group agreed that Hopkins should be persuaded to devote all his time to research, instead of wasting his genius in running what they saw as an over-large and disorderly department. Having a particularly low opinion of Hopkins' administrative and organisational abilities, they made concerted efforts to relieve Hopkins' of his administrative duties. Hopkins resisted these attempts.

In 1927, Hill and Elliott, with Fletcher's blessing, conceived of a plan to secure a research professorship for Hopkins and bring back another Cambridge man to head and manage the Dunn Institute. They opted for George Barger, Professor of Medical Chemistry in Edinburgh, who was willing. Hopkins, too, seemed favourably disposed to the plan. It proved less straightforward to secure the funds required for Hopkins. The most promising approach Hill and Elliott made was to the Beit Trustees, for which they canvassed powerful support from Anderson, Dale and other eminent scientists such as Charles Sherrington. Because of the unusual nature of the scheme, however, the Trustees demanded unanimous approval from its Advisory Board. In the event, the Board was split, and the 'Hopkins relief bill', as it was nicknamed, was rejected. Hill suspected that Hopkins himself had been reluctant to push for change, as a result of special pleading on the part of his 'young friends', who probably feared that 'a more determined and less pliable head of department would be appointed, so that it would not be a "free for all" there'.

THE TRADITION DEFENDED

Hopkins resisted any further attempt to change the administration of his department, having good reason to believe that the real target was the entire research ethos of his school. In response to this challenge, he adopted two strategies. One was simply to refuse to retire, which was his statutory right. The other was to keep on making 'propaganda', as he called it, for his particular vision of biochemistry as a fundamental, unifying science of life. In numerous public lectures, he presented bio-

chemistry as the science that is uniquely suited to producing an intellectual synthesis of biology and chemistry. Even his last major public address, the Linacre Lecture of 1938, had the highly significant title of 'Biological thought and chemical thought: a plea for unification'.

Hopkins frequently returned to themes he had already stressed in his famous lecture to the British Association in 1913: the integrated dynamics of the living cell, the simplicity of chemical events underlying biological functions, and the need for fundamental research. He introduced significant new themes, too. In particular, Hopkins increasingly voiced his conviction that investigations of cellular dynamics would reveal an underlying unity of biochemical processes across the living world. Comparative studies were essential to achieve this end: only insights into the basic biochemical processes in a wide variety of tissues and organisms, from bacteria to humans, would make it possible to identify the common biochemical patterns that underlie the great diversity and specificities of biochemical events. By giving his lectures substance with illustrations drawn to a large extent from his own department's research, Hopkins implied that this search for common patterns was exactly what motivated, and unified, his school's apparently disconnected research efforts.

Hopkins' ceaseless propaganda makes sense in the light of the attempts to 'unseat' him and change the directions of his school's research. Internally, his institute was flourishing, its teaching programmes were well entrenched in the University, and his colleagues had been putting his principles into practice with unquestionable commitment and enthusiasm. Although Hopkins' school was seen as exemplary by most of those who worked in or visited the department, it remained highly atypical. In the interwar period, the majority of research and teaching in biochemistry remained closely allied to physiology and medicine. Hopkins promoted a different route for biochemistry during this crucial period when it was being consolidated as an academic discipline, but one that many of his peers in the British scientific establishment judged too unorthodox.

When Hopkins finally retired in 1943, he was succeeded in the

Chair of Biochemistry by A.C. Chibnall, from Imperial College, London. In the event, Chibnall did not enjoy managing the large Dunn Institute, being much happier in the laboratory. While preparations were in progress for the First International Congress of Biochemistry, held in Cambridge in August 1949, Chibnall made it known that he intended to resign his Chair once he had completed his task as President of the congress.

Just at this time, Hopkins' former colleagues made a concerted attempt to bolster the Hopkins tradition as they had experienced it, presenting Hopkins' school as a model of successful biochemistry. Following Hopkins' death in 1947, they prepared a commemoration volume entitled *Hopkins & Biochemistry, 1861–1947*, edited by Joseph Needham and Ernest Baldwin. Published in 1949, it was presented to each of the more than 1700 congress participants. Besides a generous Foreword by Chibnall, the book contains an autobiographical piece written by Hopkins in 1937, excerpts from Hopkins' scientific papers with in-depth commentaries, and essays about Hopkins by Marjory Stephenson and by Joseph and Dorothy Needham. Here Hopkins' leadership is presented as incontestably inspiring and enlightened, and his school as a mecca of biochemistry. Selections from *Brighter Biochemistry* and photographs of Hopkins and colleagues serve to enhance the impression of a cohesive 'scientific family'.

Above all, the volume brings together fifteen of Hopkins' public addresses, some of them previously unpublished. Hopkins had never written a book on biochemistry and his published lectures were scattered through the literature. Together, the lectures selected by his colleagues make a powerful statement of Hopkins' ambitions for biochemistry, giving substance to Stephenson's evaluation that Hopkins 'alone among his contemporaries succeeded in *formulating* the subject'.

Not even this passionately positive portrayal of the Hopkins tradition helped secure the Chair of Biochemistry for an internal candidate when Chibnall's successor was elected. Neither Joseph Needham

nor Malcolm Dixon, both of whom had been closely associated with Hopkins, was appointed. For Hopkins' colleagues, the era that had shaped them now definitely came to an end. Of the three prime movers behind *Hopkins & Biochemistry*, Marjory Stephenson died of cancer in 1948, Ernest Baldwin moved to the Chair of Biochemistry at University College London, and Joseph Needham moved out of biochemistry altogether to devote his time to historical studies of the science and civilisation of China. (He did, however, retain his Readership in Biochemistry until 1966.) Yet the volume they put together has exerted a lasting influence as a tribute to the Hopkins tradition.

With the enormous expansion of funding of basic research in the life sciences after the Second World War, Hopkins' earlier articulations of biochemistry as a fundamental science of life gained new resonance. From the 1950s, Hopkins was regularly called 'the father of British biochemistry' and, retrospectively, the scope and objectives of his school's research in the interwar period came to be seen as exemplary. It has largely been forgotten that the conduct and organisation of research in this model school was criticised severely at the time, within and beyond the University of Cambridge. In view of the model status that interwar Cambridge biochemistry has nevertheless acquired, it seems fair to conclude that the University owes rather more to the efforts of Hopkins and his 'young friends' than their eminent critics gave them credit for.

Further reading

E. Baldwin. 1981. Hopkins, Frederick Gowland. In *Dictionary of Scientific Biography*, ed. C.C. Gillispie, vol. VI, pp. 498–502.

H.H. Dale. 1948–9. Frederick Gowland Hopkins. *Obituary Notices of Fellows of the Royal Society,* **6,** 115–45.

H. Kamminga and M.W. Weatherall. 1996. The Making of a Biochemist. I: Frederick Gowland Hopkins' Construction of Dynamic Biochemistry. *Medical History,* **40,** 269–92.

R.E. Kohler. 1982. *From Medical Chemistry to Biochemistry*. Cambridge University Press, chapters 3 and 4.

J. Needham and E. Baldwin (eds). 1949. *Hopkins & Biochemistry, 1861–1947*. Cambridge: Heffer.

M.W. Weatherall and H. Kamminga. 1996. The Making of a Biochemist. II: The Construction of Frederick Gowland Hopkins' Reputation. *Medical History*, **40**, 415–36.

Sir Frederick Gowland Hopkins (1861–1947) was born in Eastbourne, East Sussex. Brought up by his widowed mother, he was educated at the City of London School, where he excelled at chemistry. After working as an insurance clerk for six months, Hopkins was articled to a consulting analyst. His outstanding performance in the Associateship Examination of the Institute of Chemistry brought him to the notice of Sir Thomas Stevenson, Home Office Analyst, who engaged Hopkins as his assistant. Hopkins entered Guy's Hospital Medical School in 1888, qualifying in 1894, and also obtained a B.Sc. in chemistry from the University of London in 1890. He worked as a demonstrator in physiology at Guy's Hospital from 1894 to 1898.

Hopkins was appointed Lecturer in Chemical Physiology at the University of Cambridge in 1898 and promoted to Reader in 1902, with financial backing from Emmanuel College. In 1910, Trinity College elected him Fellow and Praelector in Biochemistry. In 1912 he became co-director of the Institute of Animal Nutrition, Cambridge. He was elected Professor of Biochemistry at Cambridge in 1914, and his title was renamed as Sir William Dunn Professor of Biochemistry in 1921. He retired in 1943.

Hopkins was elected Fellow of the Royal Society in 1905 and served as its President from 1931 to 1935. He was awarded the Royal Society's Copley Medal in 1926 and shared the Nobel Prize for Physiology or Medicine in 1929. He was knighted in 1925 and received the Order of Merit in 1935.

Hopkins married Jessie Anne Stevens in 1898. They had two daughters and a son.

Harmke Kamminga is a Wellcome Trust Senior Research Fellow in History of Medicine, working in the Department of History and Philosophy of Science, University of Cambridge. Over the last twenty years, she has published on a wide range of subjects in the history of biological and biomedical sciences. Among recent works, she co-edited (with Andrew Cunningham) *The Science and Culture of Nutrition, 1840–1940* (Rodopi, 1995) and (with Soraya de Chadarevian) *Molecularizing Biology and Medicine: New Practices and Alliances, 1910s-1970s* (Harwood, 1998). Her current research is centred on the history of biochemistry and the history of vitamins.

13 Charles Sherrington, E.D. Adrian, and Henry Dale: The Cambridge Physiological Laboratory and the physiology of the nervous system

E.M. TANSEY

The Cambridge Physiological Laboratory was formally established in 1883 when Michael Foster (1836–1907), then Praelector in Physiology at Trinity College Cambridge, accepted Cambridge University's first Chair of Physiology. From this laboratory emerged several key scientists in the study of the nervous system, principal amongst them being Foster's colleagues J.N. Langley (1852–1925) and Walter Gaskell (1847–1914). Their pupils Charles Sherrington (1857–1952), Henry Dale (1875–1968), and E.D. Adrian (1889–1977) all won Nobel Prizes for elucidating basic mechanisms of the nervous system, each of them making major contributions to modern understanding of the functional mechanisms of the nervous system. In turn, in 1963, two of Adrian's own pupils Alan Hodgkin (1914–1998) and Andrew Huxley (b.1917) became Nobel laureates in 1963 for their investigations of the molecular mechanisms of neural activity. Also from the Physiological Laboratory in the earlier period came A.V. Hill (1886–1977) who won the Nobel Prize in 1922 for his work on heat generation by nerve and muscle. The physiological research work that Sherrington, Dale, and Adrian undertook, and the lab from which they emerged, in which Langley, Gaskell, Adrian, and Hodgkin spent almost their entire professional careers, will form the main foci of this chapter.

MICHAEL FOSTER AND THE CAMBRIDGE PHYSIOLOGICAL LABORATORY: FEW APPOINTMENTS HAVE MORE PROFOUNDLY INFLUENCED THE FUTURE OF A UNIVERSITY OR A SUBJECT

Michael Foster trained under William Sharpey, the 'father of British physiology', at University College London, becoming Professor of Practical Physiology there in 1867. In 1870, persuaded, according to one history of the Cambridge Medical School by the novelist George Eliot (Mary Ann Evans) and her common-law husband, the writer and amateur scientist George Henry Lewes, he accepted Trinity College's invitation to become its inaugural Praelector in Physiology. Trinity College made several important contributions to the advancement of late nineteenth century science, and its support for physiology in particular has been recognised by previous authors, especially Gerald Geison, who has examined the development of the Cambridge 'school'. Of particular significance was the proposal in the late 1850s by the Statutory Commission on Cambridge University that there should be more scientific instruction at Trinity College, and although immediate response was somewhat reluctant, powerful College personalities became involved over the following years in stratagems to promote the status of science, including Henry Sidgwick, J. Willis Clark, and Coutts Trotter.

Manoeuvrings within Trinity College reflected change and liberal revival in Cambridge, within a general climate of increased access to higher education, especially for certain classes and religious adherents. The University Tests Act of 1871, by removing theological requirements for admission to Fellowships, opened up the possibility of Oxbridge University careers to nonconformists. Contemporaneously, a campaign for the promotion of science, frequently referred to as the 'Endowment of Research Movement', spread across the country, gaining considerable momentum towards the end of the nineteenth century. There was clear recognition that the availability of facilities (buildings, equipment, and supplies) and positions (academic posts, research Fellowships, and training studentships) were of critical impor-

tance in the prosecution of the new, laboratory-based, experimental sciences. Michael Foster's appointment to Trinity College epitomises this movement.

Foster used his new opportunities to promote physiology in several ways: he attracted pupils and colleagues to Cambridge; his *Textbook of Physiology* (first edition, 1877) reached further afield; in 1873, a collection of *Studies from the Physiological Laboratory in the University of Cambridge* appeared that heralded the inauguration, in 1878, of the *Journal of Physiology*; in 1875 he presented critical evidence on the validity and necessity of animal experimentation to the Royal Commission investigating the practice; and in the following year was the prime mover in establishing the Physiological Society. These and other commitments illustrate his pivotal role in the development of modern British physiology, a position substantially reinforced in 1883 when he became Cambridge's first Professor of Physiology.

Foster's pull on younger physiologists was obvious from the beginning of his association with Cambridge: he persuaded Newell Martin, later Professor of Physiology at Johns Hopkins University, to move from University College London, and directly inspired, amongst others, later distinguished scientists such as W.H. Gaskell, F.M. Balfour, A.G. Dew Smith, Sheridan Lea, J.N. Langley, H.K. Anderson, C.S. Sherrington, and W.B. Hardy. The success of the Cambridge Laboratory and the shaping and success of its research did not, and could not, of course, depend solely on one man, and the research reputation that the Physiological Laboratory rapidly acquired also owed much to two of Foster's staff, Langley and Gaskell. Foster's outside duties, as Biological Secretary and Council member of the Royal Society, and his service on numerous Royal Commissions, meant that it was Langley and Gaskell who most profoundly influenced the younger physiologists who made their names and reputations in the new century. Their expertise, especially in the study of the anatomy, physiology, pharmacology, and evolution of the nervous system, directly influenced their pupils Charles Sherrington, Henry Dale, and E.D. Adrian.

J.N. Langley read medicine at Cambridge, and spent his entire professional life as a member of the Physiological Laboratory and a Fellow of Trinity College. In 1903 he was appointed as Foster's deputy; succeeding to the Chair in 1907, he remained as professor until his death in 1925. In the 1880s Langley began his major systematic study of the physiology and pharmacology of what was variously known then as the 'visceral' or 'involuntary' nervous system. After discussions with his Cambridge classical colleague, Professor Jebb, Langley suggested an alternative word, 'autonomic' for this component of the nervous system. His experiments significantly extended and complemented those of his colleague Walter Gaskell, and were of considerable importance for the study of neural functioning. Gaskell's work ranged widely, from cardiac function and innervation, to detailed studies of the sympathetic nervous system, and to speculations on vertebrate evolution. His laboratory studies were predominantly anatomical, and his detailed histological examinations contributed greatly to the unravelling of the component parts of the autonomic nervous system and provided the morphological base of our modern knowledge of the autonomic nervous system. So much so, that many years later an eminent neurologist Walter Langdon-Brown (1870–1946) declared that 'to read an account of the [autonomic nervous] system before Gaskell is like reading an account of the circulation before Harvey'. Gaskell, like many others, felt an enormous debt of gratitude to Michael Foster, as he demonstrated in his obituary, 'by his earnestness, his lovable charm of persuasion, his entire freedom from any thought of monetary gain, or any kind of selfishness, the conviction was gradually borne in on his pupils that the particular line of research on which each was engaged was the one thing in life worth doing, and the only place to do it was in Cambridge by Foster's side'.

CHARLES SHERRINGTON: CAMBRIDGE AND BEYOND

One person who started 'by Foster's side' was Charles Sherrington, who began studying medicine at St Thomas' Hospital London in 1876, and, when family finances improved sufficiently to allow it, became a

scholar at Gonville and Caius College, Cambridge in 1881. He was much influenced by Foster to study physiology, and with Walter Gaskell as his tutor, took a first class in both parts I and II of the Natural Sciences Tripos, before completing his clinical training at St Thomas' Hospital. In later life Sherrington acknowledged Gaskell's subtle inspiration, sympathetic attention, and transcendent sincerity. His first scientific paper however was the result of a neuroanatomical study with J.N. Langley on a dog that had had part of its forebrain surgically removed, in an experiment to correlate behavioural deficits seen in the dog during its lifetime, with morphological changes in its brain. The paper was read to the Royal Society of London and was the first of over 300 research publications that Sherrington produced over his long life.

After leaving Cambridge, Sherrington spent some months travelling and visiting laboratories in Europe, before returning to St Thomas' Hospital as its first full-time lecturer in physiology in 1887. Until then physiology had merely been the part-time interest of a member of the medical staff, but this new academic position was poorly paid, and Sherrington was also employed as Professor Superintendent of the Brown Animal Sanatory Institution in South London, and whilst there became, in 1894, one of the first people in Britain to use the new therapeutic wonder-drug of the period, diphtheria anti-toxins. However, it was whilst at St Thomas' that he began his studies of the structure and function of the nervous system. He was deeply influenced by the writings of the Spanish neuroanatomist Santiago Ramon y Cajal, who had described the individual nerve cell, or neuron, as the basic *anatomical unit* of the nervous system, in contrast to the Italian histologist Camillo Golgi who maintained that the nervous system was made of an interconnecting network of cells and fibres. Sherrington in turn described the *functional unit* of nervous activity, the reflex arc, which in its simplest form consists of just two neurons, as an input and an output neuron, the communication between which requires conduction of the neural impulse across the intervening space. Reminiscent of Langley's creation of the term 'autonomic', and after consultation with

a classical scholar from Cambridge, Sherrington coined the word 'synapse' to describe this gap between nerve cells, the word first appearing in the 1897 edition of Michael Foster's *Textbook of Physiology*. In later years it was Henry Dale and his colleagues who elucidated the mechanisms whereby specific chemicals effected communication across the synapse.

Sherrington further showed that the nervous system is composed of successive reflex arcs at higher and higher levels, and that many routine actions, such as sneezing or walking, could be analysed into their component series of reflexes. From this work he postulated the existence of 'the final common pathway' by which he meant that many different sensory inputs could, through a cascade of reflex arcs, all have one output neuron. In turn that neuron would innervate the final effector, usually a muscle cell.

Sherrington described the reciprocal innervation of antagonistic muscles, by which the activity of one set of excited muscles is integrated with another set of inhibited muscles, and contributed a second major concept to twentieth-century neuroscience, that of integration, an holistic view of nervous function, coordination and connectivity, which explains how the nervous system can receive and respond to information from the external world. These views he postulated in 1906 in *The Integrative Action of the Nervous System*, which, E.D. Adrian later commented, 'opened up an entirely new chapter in the physiology of the nervous system'. In 1932 they shared the Nobel Prize in Physiology or Medicine for their contributions to understanding the nervous system.

In 1895 Sherrington moved to Liverpool as Professor of Physiology, and in 1913 left to go to the Chair at Oxford. His intellectual interests extended beyond the nervous system: he was a keen sportsman, as a schoolboy and student he played soccer for Ipswich Town; in addition to his research he produced a textbook on practical mammalian physiology, which influenced generations of physiologists; he was a published poet and wrote on the nature and history of science; in his late fifties during the First World War he worked as an

ordinary hand in a munitions factory because he was chairman of the Industrial Fatigue Board; and was a noted bibliophile, regarded by the British Library as one of their major patrons. Like Dale and Adrian, he served as President of the Royal Society of London, his term lasting from 1920–5, and he retired in 1935, heaped with honours and distinctions, including a knighthood in 1922, and was renowned as a philosopher of the nervous system.

HENRY DALE AND CHEMICAL NEUROTRANSMISSION ACROSS THE SYNAPSE: THE GOLD THAT KEEPS ITS VALUE

Michael Foster held firm views on the importance of teaching physiology, believing the best educational methods were demonstrations and practical exercises. This he put into notable practice on 25 November 1893, when he gave a talk and demonstration on 'Nerves' to the Natural History and Science Society of the Leys School in Cambridge. The schoolboy editor of the school magazine wrote up the meeting vivaciously, describing how stimulation of different nerves either slowed or speeded up the beating of a frog's heart, and how the application of drugs could mimic those effects. That child was Henry Dale who, more than forty years later, was to share the Nobel Prize for explaining the physiological basis of some of those effects.

Henry Dale later recalled how, to his immense disappointment, as an undergraduate, he found Foster, who taught the 'Elementary Course', boring and repetitive, but that Gaskell and Langley, who directed an advanced physiology course, gave him more satisfaction and encouragement, both in their formal lectures and by their informal example. Langley gave the entire lecture series (elementary and advanced) on the central nervous system, and Dale commended him particularly as 'an imperturbably accurate observer, eager for an accurate account of observed facts and impatient of speculative theory'. Also recalling his weekly essays and Langley's dry critical comments, Dale acknowledged a 'hard, but a wholesome discipline' that lasted for the rest of his career. Gaskell he considered 'certainly the most stimulating teacher of advanced students that I have ever encountered. He

gave the hearers an inspiring vision of research, and of course physiological research, as an exciting intellectual adventure.'

Dale's own exciting intellectual adventure was slow to get started. On completion of his Tripos, with first class honours, he was encouraged by members of the Physiological Laboratory to continue but financial problems haunted his early efforts to do so. For two years he shared the Coutts Trotter studentship of Trinity College with the Hon R.J. Strutt, son of the 3rd Lord Rayleigh, but when the two men competed for a Fellowship in 1900, a tied vote resulted in the Master, Montagu Butler, apparently deciding in favour of Strutt. Returning to London to qualify in medicine at St Bartholomew's Hospital, Dale had a second chance at a research career in 1902 when, supported by Michael Foster, he won the George Henry Lewes Studentship to work with Ernest Starling and William Bayliss in the Physiology Department of University College. In 1904 a most unusual opportunity presented itself: the pharmaceutical manufacturer Henry Wellcome, the proprietor of Burroughs, Wellcome & Co, offered Dale the position of staff pharmacologist at his private research laboratories. Deeply suspicious of commercial associations, Cambridge friends advised Dale 'not to sell his scientific birthright for a mess of commercial pottage'. Intrigued, however, by the exceptional opportunities offered, and with little other chance of staying in scientific research, Dale accepted the post. He remained there for ten years, leaving in 1914 to enter the service of the newly created Medical Research Committee (later Council, the MRC) as a member of staff of their embryonic National Institute for Medical Research (NIMR). After the disruptions of the First World War, the NIMR acquired a permanent building in North London, and Dale was appointed its first Director in 1928, which he held until retirement in 1942. Like the other two subjects of this chapter, he received numerous civilian and professional honours, including a knighthood awarded in 1932, his old boss Henry Wellcome being in the same honours list for the same award. Exactly twenty years after Sherrington had done so, he served as President of the Royal Society (1940–45), which at that period involved considerable respon-

sibilities, including membership of the Scientific Advisory Committee to the War Cabinet.

The Wellcome Physiological Research Laboratories had originally been established in 1894 to produce diphtheria anti-toxin, but by 1904, when Dale joined, Henry Wellcome was keen that his staff undertake original research that might, or might not, be of direct benefit to his commercial company. Dale thrived in this environment: he worked in well-equipped laboratories, with no teaching duties, and with an almost free hand as to which research problems he followed. When he arrived, Wellcome had suggested that as he had no immediate research project, Dale might like to examine the pharmacology of a naturally occurring fungus, ergot of rye. At the time, ergot was produced commercially to prevent post-partum haemorrhage, and Wellcome was actually hoping for a 'cleaner', less-contaminated preparation than that sold by rival drug companies. Dale undertook several studies of the effects of pharmacologically active extracts of ergot on elements of the autonomic nervous system, and between 1904 and 1914, when he left the Wellcome Laboratories, Dale and colleagues discovered several important biologically active chemicals, all initially extracted from ergot, including acetylcholine, histamine, and tyramine. He also extracted a chemical that inhibited the natural action of adrenaline, a forerunner of modern beta-blocker drugs. From these studies stemmed all his subsequent work, on the isolation, identification, and detailed examination of the physiological roles of a wide range of endogenous chemicals.

During his period at the NIMR, much of his earlier work came to fruition, especially that on the role of acetylcholine. With a number of colleagues, including Wilhelm Feldberg and Marthe Vogt, both refugees from Hitler's Germany, G.L. Brown, and John Gaddum, Dale showed that acetylcholine was a normal constituent of the mammalian nervous system, which served at some synapses, the junctions in the nervous system named by Sherrington, as a chemical messenger, passed from one nerve cell to another across the gap between them. For his experimental work on chemical neurotransmission he shared the

Nobel Prize in 1936 with the Austrian pharmacologist, Otto Loewi. Another important contribution that emerged from these experimental studies was the creation of words to describe the new phenomena he discovered, following in the footsteps of both Langley and Sherrington. To describe chemicals that mimicked the actions of the sympathetic nervous system, Dale invented the word 'sympathomimetic', and to delineate the effects of parts of the nervous system that might use acetylcholine or adrenaline-like substance as their neurotransmitter, the words 'cholinergic', 'adrenergic'.

Dale always refused to develop an explicit theoretical concept to account for the results he obtained, quoting his Cambridge teacher 'Make accurate observations and get the facts', he [Langley] would say; 'if you do that the theory ought to make itself'. In an appreciation of Dale, E.D. Adrian emphasised the point further:

> no one can have seen Dale at work in the laboratory or listened to his communications to the Physiological Society without realizing that it is the evidence which is all-important. Dale has never let his theories take charge: often enough his evidence has led to elaborate theorizing by others – about the role of histamine for instance, or about humoral transmitters in the nerve fibre. When such new horizons are opened it is hard to keep to the solid and familiar ground, but Dale has been more concerned to apply the brake than to be the first in the gold rush. The gold he has found will keep its value.

E.D. ADRIAN, BARON ADRIAN OF CAMBRIDGE

Edgar Douglas Adrian went up to Trinity College, Cambridge in 1908, as a Major Scholar in Natural Science, achieved spectacularly high marks in his part I, concentrating on physiology for his part II, to gain a first class honours degree in 1911. In addition to his studies he engaged in a number of sporting activities and regularly attended, as had Henry Dale before him, the Cambridge University Natural Science Club. Unlike Dale, he achieved almost immediate academic success in Cambridge: he was elected a Fellow of Trinity in 1913, shortly before leaving

Cambridge to complete his clinical studies at St. Bartholomew's Hospital. After war work as a neurologist, he returned to Cambridge, where he spent the rest of his lengthy career, devoted to studying the electrical mechanisms by which nerve fibres transmit information. He became a Foulerton Research Professor of the Royal Society in 1929, succeeded Sir Joseph Barcroft as Professor of Physiology in 1937, retaining the Chair until 1951 when he became Master of Trinity. Thirty years after Sherrington, and ten years after Dale's tenure, Adrian too served as President of the Royal Society (1950–5), and in later years two of his pupils and Fellow Nobel laureates, Alan Hodgkin (1970–5) and Andrew Huxley (1980–5), also became Presidents. Adrian became Vice-Chancellor of Cambridge University from 1957, and ten years later accepted the duties of Chancellor. He retired from the Mastership of Trinity in 1965, but following his wife's death shortly afterwards, moved back into College, where he remained until his death in 1977.

As an undergraduate, Adrian's Director of Studies at Trinity was the neurophysiologist Keith Lucas (1879–1916) whose work on the conduction of the nervous impulse was a major influence on his own career. When he resumed his academic career in 1919 Adrian inherited Lucas' laboratory, became his mentor's literary executor, and began a study of the recovery of nerve and muscle fibres after excitation. Of considerable interest to physiologists at this time was the problem of amplifying the very small electrical signals that they recorded and Adrian was successful in developing new techniques to record nervous activity. In 1925 he reported on the use of a triode valve amplifier, then being used in the young telegraphic industry, to magnify the responses he recorded by a factor of nearly 2000. He was thus able to record the activity of sensory and motor nerve fibres in a wide variety of anaesthetised animals and in humans, and later developed a particularly sensitive technique of recording from single, isolated nerve fibres, in which he showed that there is only one *kind* of nervous impulse and that all neural information is conveyed by variations in the frequency at which those impulses are transmitted, the so-called 'frequency code', an essential characteristic of nerves. This is of fundamental significance

in understanding the functioning of the nervous system. Either alone or in collaboration, Adrian also investigated the ways in which a variety of sense organs in many different species function; the comparative approach characterising much of his work. He mapped out the cortical representations in the brain of pressure and touch receptors in many different animals, and frequently illustrated his papers with his own drawings. His artistic nature and his sense of humour had been demonstrated many years earlier when, as an undergraduate, he participated in a spoof 'Post-Impressionist' exhibition in Cambridge, exhibiting several of his own works including a unique 'Self-portrait'. In 1934 Adrian and Bryan Matthews confirmed and extended the observations of the Swiss neurologist Hans Berger, that electrical potentials could be recorded from conscious patients. They devised methods to record and analyse the gross electrical activity of the brain, and electroencepholography (EEG) has become an important clinical tool in the study of brain function in health and disease, although the precise significance of the recordings remains an area of debate. Adrian shared the 1932 Nobel Prize for Physiology or Medicine with Charles Sherrington 'for their discoveries regarding the function of the neurone', and subsequently received numerous honours, including one of the last hereditary peerages to be created in 1955. Despite his numerous academic and other commitments he continued to work, usually on his own, until his late sixties, when a disastrous flood wrecked his basement laboratory that had been untouched by cleaner or decorator for over forty years. The resultant mess and damage to his equipment effectively terminated Adrian's career as an experimentalist, although he continued to lecture and write on the nervous system until his death.

Like his own teachers in Cambridge, his influence was frequently acknowledged, as in this posthumous tribute:

> Lord Adrian's influence upon the sensory sciences was great, not only in terms of his contribution to knowledge itself but through the influence he exerted upon numerous young scientists who spent weeks or years at the Cambridge laboratory and who later returned

to their homelands and colleagues with the seeds of vigorous research and quantitative inquiry firmly planted.

All three men made major advances in understanding the mechanisms of the nervous system: Adrian particularly by his work on the way the individual nerve cells function; Sherrington for examining the processes by which the actions of individual nerve cells integrate into coordinated actions; and Dale for unravelling the methods by which individual nerve cells communicate with each other, by means of chemical neurotransmission. Despite their professional eminence and the fundamental importance of their work, Sherrington, Dale, and Adrian are no longer well-known names. The twentieth century has seen some cultural ambivalence towards scientific achievements, as recognised by a contemporary of theirs, C.P. Snow, in his lecture 'The two cultures'. 'I remember G.H. Hardy once remarking to me in mild puzzlement, some time in the 1930s: "Have you noticed how the word 'intellectual' is used nowadays? There seems to be a new definition which certainly doesn't include Rutherford or Eddington or Dirac or Adrian or me. It does seem rather odd, don't y'know."'

Further reading

Lord Adrian. 1955. Sir Henry Dale's contribution to physiology. *British Medical Journal*, **1**, 1355–6.

S.J. Curtis. 1968. *History of Education in Great Britain*. London: University Tutorial Press Ltd.

H.H. Dale. 1953. *Adventures in Physiology*. London: Pergamon Press. Reprinted 1965 by the Wellcome Trust, London.

J.C. Eccles and W.C. Gibson. 1979. *Sherrington: His Life and Thought*. Heidelberg: Springer International.

W.S. Feldberg. 1970. Henry Hallett Dale 1875–1968. *Biographical Memoirs of Fellows of the Royal Society London*, **16**, 77–174.

G.L. Geison. 1978. *Michael Foster and the Cambridge School of Physiology. The Scientific Enterprise in Late Victorian Society*. Princeton, NJ: Princeton University Press.

Alan Hodgkin. 1979. Edgar Douglas Adrian, Baron Adrian of Cambridge, 1889–1977 *Biographical Memoirs of Fellows of the Royal Society*, **25**, 1–73.

H.D. Rolleston. 1932. *The Cambridge Medical School. A Biographical History*. Cambridge: Cambridge University Press.

S. Rothblatt. 1968. *The Revolution of the Dons: Cambridge and Society in Victorian England*. London: Faber & Faber.

C.P. Snow. 1959. *The Two Cultures and the Scientific Revolution. The Rede Lectures*. Cambridge: Cambridge University Press.

Edgar Douglas Adrian, 1st Baron Adrian of Cambridge (1889–1977). Neurophysiologist born in London, who trained in the Physiological Laboratory, Cambridge, and became a Fellow of Trinity College. After working as a neurologist during the First World War he returned to Cambridge and began his life's research on the nervous system, studying in particular the electrical impulses along nerve fibres, and the mechanisms by which nerve transmit information. He demonstrated, from recordings in both motor and sensory nerves, that there is only one kind of nervous impulse, information being conveyed by variations in the frequency of impulse transmission, thus defining an essential characteristic of nerves, the 'frequency code'. Adrian also studied the ways in which the sense organs of many different kinds of animals detect and transmit sensation, and was involved in the development of electro-encephalography as a clinical diagnostic tool. He became Professor of Physiology in Cambridge (1937–51), Master of Trinity (1951–65), Vice-Chancellor of Cambridge University (1957–67), and subsequently Chancellor (1968–75). He was elected a Fellow of the Royal Society in 1923, served as Foreign Secretary (1945–50) and as President (1950–5). He shared the Nobel Prize in Physiology or Medicine for 1932 with Sir Charles Sherrington, became a member of the Order of Merit in 1942, and received a hereditary peerage in 1955.

Sir Charles Scott Sherrington (1857–1952). Neurophysiologist born in London and trained in the Physiological Laboratory, Cambridge and St Thomas' Hospital, London. After spending some time in European laboratories, he became the first full-time professional Lecturer in Physiology at St Thomas' (1887–95) where he began his life's work on the structure and function of the nervous system. He defined the functional unit of the nervous system as being the reflex arc, and recognised that the nervous system is composed of successive reflex arcs at higher and higher levels. He suggested that there was a 'final common pathway' by which many different sensory inputs could, through a series or different reflex arcs, all have one output neuron. He also proposed an holistic view of nervous function, coordination, and connectivity, which he postulated in *The Integrative Action of the Nervous System* (1906), a book which consitituted a significant landmark in

modern neurophysiology. He also coined the word 'synapse' to describe the junction between nerve cells, and in 1932 shared the Nobel Prize for Physiology or Medicine with E.D. Adrian. He became Professor of Physiology at Liverpool (1895–1913) and then in Oxford (1913–35), was elected a Fellow of the Royal Society in 1893 and served as President (1920–5). He was knighted in 1922 and became a member of the Order of Merit in 1924. In addition to his scientific work, he published poetry and was a great bibliophile, contributing numerous valuable volumes to the British Museum, of which he was a Trustee.

Sir Henry Hallett Dale (1875–1968). Physiologist and pharmacologist born in London, and trained in the Physiological Laboratory, Cambridge. After brief spells of research with John Langley in Cambridge and Ernest Starling and William Bayliss at University College, he became Staff Pharmacologist at the Wellcome Physiological Research Laboratories (1904–14). His work was principally on the physiology and pharmacology of naturally occurring chemicals, much deriving from experimental work on the fungus ergot of rye. He discovered a chemical that reversed the effect of adrenaline, a crude forerunner of modern beta-blocker drugs, and examined numerous chemicals that mimicked the effects of normal sympathetic nervous system action. He discovered histamine, tyramine, and acetylcholine, all from extracts of ergot, and his later work showed that acetylcholine occurred naturally in animals, and played an important role in the transmission of nerve impulses across the synapse. For this work he shared the 1936 Nobel Prize in Physiology or Medicine with the Austrian pharmacologist Otto Loewi. Dale also did pioneer work in endocrinology, and closely associated with his research were his wider concerns with the standardisation of drugs, and he served on many advisory and regulatory committees at both national and international levels. In 1914 he joined the National Institute for Medical Research (NIMR) and became its first Director (1928–42). He was elected a Fellow of the Royal Society in 1914, and served as Biological Secretary (1925–35) and President (1940–5). He was knighted in 1932 and created a member of the Order of Merit in 1944.

E.M. Tansey studied the neurochemistry of the *Octopus* for her Ph.D. After working in several neuroscience labs she retrained as a historian, and was awarded a Ph.D. in the history of medicine in 1990. She is historian of modern medical sciences at the Wellcome Centre for the History of Medicine at University College London. She has published widely on the history of twentieth-century medical sciences, and is co-editor of the *Wellcome Witnesses to Twentieth Century Medicine* series. She is currently working on a history of the British pharmaceutical industry, and completing a study of the technician in medical laboratories.

14 Hardy and Littlewood
ROBIN J. WILSON

THE HARDY-LITTLEWOOD PARTNERSHIP

The mathematical collaboration of Godfrey Harold Hardy and John Edensor Littlewood is the most remarkable and successful partnership in mathematical history. From before the First World War until Hardy's death in 1947 these mathematical giants produced around one hundred joint papers of enormous influence covering a wide range of topics in pure mathematics. Whereas many other mathematicians have collaborated on a short-term basis, there are no other examples of such a long and fruitful partnership.

Hardy and Littlewood dominated the English mathematical scene for the first half of the twentieth century. Throughout the nineteenth century, mathematical life in England, especially in pure mathematics, had been dwarfed by developments on the Continent, and although Cambridge had produced some outstanding applied mathematicians, such as James Clerk Maxwell, George Gabriel Stokes, and William Thomson (Lord Kelvin), there were few pure mathematicians of world class other than Arthur Cayley in Cambridge and James Joseph Sylvester in Oxford. The situation changed with Hardy and Littlewood, who created a school of mathematical analysis unequalled throughout the world. As one contemporary colleague observed: 'Nowadays, there are only three really great English mathematicians, Hardy, Littlewood, and Hardy–Littlewood.'

As frequently happens in collaborative partnerships, the styles and personalities of the two men were very different. Both were mathematical geniuses, completely devoted to their subject, and with many interests in common. But Littlewood was probably the more original of the two, imaginative and immensely powerful and enjoying the challenge of a very difficult problem, while Hardy was the consummate

FIGURE 14.1 Hardy and Littlewood at Trinity College.
Source: Trinity College, Cambridge.

craftsman, a connoisseur of beautiful mathematical patterns and a master of stylish writing. As Littlewood once observed: 'My standard role in a joint paper was to make the logical skeleton, in shorthand – no distinction between r and r^2, 2π and 1, etc., etc. But when I said "Lemma 17" it stayed Lemma 17.' Hardy, who considered Littlewood as the finest mathematician he had ever known and the more creative partner in the collaboration, always wrote the final draft of their joint papers.

Although they both lived in Trinity College for part of their collaboration, their joint work was conducted largely through letters and postcards. According to the Danish mathematician Harald Bohr, they initially had misgivings about collaborating, fearing that it might encroach on their personal freedom, so vitally important to them. As a safety measure, they formulated four 'axioms' for their collaboration. These may be briefly stated as follows:

1. When one wrote to the other, it was completely indifferent whether what they wrote was right or wrong.
2. When one received a letter from the other, he was under no obligation whatsoever to read it, let alone to answer it.
3. Although it did not matter if they both thought about the same detail, still, it was preferable that they should not do so.
4. It was quite indifferent if one of them had not contributed the least bit to the contents of a paper under their common name.

As Bohr remarked: 'Seldom – or never – was such an important and harmonious collaboration founded on such apparently negative axioms.'

Even so, it is only to be expected that two such strong personalities would occasionally come into conflict. In 1919 Hardy decided to apply for the vacant Savilian Chair of Geometry at Oxford University, and in a letter to Bertrand Russell outlining his reasons, wrote:

> I wish you could find some tactful way of stirring up Littlewood to do a little writing. Heaven knows I am conscious of my huge debt to him. But the situation which is gradually stereotyping itself is very

trying for me. It is that, in our collaboration, he will contribute ideas and ideas *only*: and that *all* the tedious part of the work has to be done by me. If I don't, it simply isn't done, and nothing would ever get published . . . At the moment I am committed to write out two joint papers for publication in Germany, inside about two months. And I can get absolutely no help from him at all: not even an enquiry as to how I am getting on! The effect on morale is most disheartening.

The difficulty was evidently short-lived, as Hardy and Littlewood produced some of their finest joint work in the years immediately following the writing of this letter.

HARDY'S EARLY YEARS

G.H. Hardy was born in Cranleigh, in Surrey, on 7 February 1877, the elder of two children of Isaac and Sophia Hardy. His father was a schoolteacher at Cranleigh School and his mother had been Senior Mistress at the Lincoln Training College. Both parents were very able and mathematically minded, but for financial reasons had not studied at university.

Hardy and his sister had an enlightened upbringing in a typical Victorian household. They were encouraged to read good literature and to discover things for themselves. He had a good all-round ability with a precocious interest in numbers – at the age of two he asked his parents to show him how to write numbers up to several millions. Hardy attended Cranleigh School until he was thirteen years old, and passed his examinations with distinctions in mathematics, Latin and drawing. He then won a scholarship to Winchester College where much of his instruction was through individual coaching rather than in class. Although he appreciated his education at Winchester, he found life there too Spartan and at one stage became very ill. Although originally intending to proceed to New College, Oxford, like many Wykehamists, a novel by Alan St Aubyn (Frances Marshall) entitled *Fellow of Trinity* inspired him to emulate its hero and head instead for Trinity College, Cambridge.

Hardy arrived in Cambridge in October 1896 and started to train for Part I of the Mathematical Tripos. This ridiculous examination was undoubtedly the most severe mathematical test anywhere, consisting of several ten-hour days of demanding problems answered against time. The candidates were trained by coaches who drilled them mercilessly, and the results were ranked in merit order – Senior Wrangler, Second Wrangler, and so on. Hardy quickly grew to detest the Tripos system and was much irritated by his own coach Dr Webb, a skilled producer of Senior Wranglers. Eventually his Director of Studies sent him to the applied mathematician Augustus Love, who inspired Hardy to read Camille Jordan's *Cours d'Analyse*. As Hardy later recalled:

> My eyes were first opened by Professor Love, who taught me for a few terms and gave me my first serious conception of analysis ... I shall never forget the astonishment with which I read that remarkable work, the first inspiration for so many mathematicians of my generation, and learnt for the first time as I read it what mathematics really meant. From that time onwards I was in my way a real mathematician, with sound mathematical ambitions and a genuine passion for mathematics.

Hardy was Fourth Wrangler in 1898, and took Part II of the Tripos examination in 1900, being placed in the first division of the First Class. He was elected to a prize fellowship at Trinity College, and in 1901 won the coveted Smith's prize (jointly with the physicist James Jeans) and a five-year prize fellowship.

It was around this time that he started to publish his mathematical discoveries. His first paper, on definite integrals, was published in the *Messenger of Mathematics* in 1900, and he followed it with many more research papers over the next ten years; the theory of integration became a constant theme and he eventually wrote no fewer than sixty-nine papers on the subject over a period of thirty years.

In 1905 Hardy wrote the first of his four Cambridge Mathematical Tracts, a much-admired work on *The Integration of Functions of a Single Variable*, and followed it with *Orders of Infinity*

(1910) and (with M. Riesz) *The General Theory of Dirichlet's series* (1915). His most celebrated book, *A Course in Pure Mathematics*, was published in 1908; a model of clarity, it presented elementary analysis to students in a rigorous yet accessible way, and through its numerous editions over many years came to have an enormous impact on English analysis.

Also in 1908 Hardy entered the world of genetics. There had been discussion about the proportions in which dominant and recessive Mendelian characters are transmitted in a large mixed population. Hardy produced a solution involving only simple algebra, and sent it to *Science*; 'Hardy's law' has since proved to be of great importance in the study of blood groups.

From 1905 to 1910 Hardy served his first term on the Council of the London Mathematical Society, which he had joined in 1901. This was the beginning of a long involvement with this and other scholarly societies. In 1910 he was elected a Fellow of the Royal Society.

In 1906 his prize fellowship was converted into a college lectureship which he held for thirteen years. This required him to lecture for six hours per week, and he usually gave two courses, one on elementary analysis and the other on the theory of functions. It was during this period, around 1912, that he started to work with Littlewood.

LITTLEWOOD'S EARLY YEARS

J. E. Littlewood was born in Rochester, in Kent, on 9 June 1885, the eldest son of Edward and Sylvia Littlewood. His father, Ninth Wrangler in the 1882 Cambridge Mathematical Tripos, was offered a Fellowship at Magdalene College but turned it down in order to become headmaster of a new school near Cape Town, in South Africa.

Littlewood lived in South Africa from 1892 to 1900, enjoying its beauty but learning little from the school or from tuition at Cape Town University. He returned to England in 1900, and attended St Paul's School in London for three years, where he learned Greek ('mainly conditional sentences about crocodiles'), developed a love for music (particularly Bach, Mozart, and Beethoven), participated actively in a

variety of sports, and learned mathematics from the distinguished algebraist F.S. Macauley. Macauley's approach was to encourage his pupils to work independently or with each other, rather than to be spoon-fed with material by the teacher – an approach that Littlewood was later to adopt with his own research students, urging them to 'Try a hard problem. You may not solve it, but you will prove something else.'

Littlewood obtained a minor Entrance Scholarship to Trinity College, and came into residence in October 1903. He was trained for Part I of the Tripos by the celebrated coach R.A. Herman (once described by Hardy as 'the mildest of the most ferocious of Huns') and became Senior Wrangler, but felt that he had wasted his first two years at Cambridge spending more than half of his time solving exceedingly difficult problems against the clock. In Part II of the Tripos he was placed in Class 1, Division 1.

Littlewood started research during the long vacation of 1906, under the direction of E.W. Barnes, later Bishop of Birmingham. Barnes believed in setting very hard problems in a sink-or-swim manner, and his first problem concerned so-called integral functions of zero order. Littlewood tackled this problem and quickly struck lucky, producing a fifty-page paper in just a few months. Littlewood's next assignment from Barnes was to prove the Riemann hypothesis, a notoriously difficult problem that remains unsettled to this day, relating to the distribution of prime numbers. Littlewood, who relished a challenge, managed to obtain some worthwhile results relating to this problem and wrote them up as a dissertation for a Trinity junior fellowship. The dissertation was well received, but the 1907 fellowship went to a classicist at his final attempt and Littlewood was awarded it the following year. Meanwhile, he won a Smith's prize, and was offered a three-year lectureship at Manchester University which he accepted, feeling that he needed a break from Cambridge. Unfortunately, the workload in Manchester was exceedingly heavy, and Littlewood came to regret his time there.

In 1910 Littlewood returned to Trinity as a college lecturer, succeeding Alfred North Whitehead whose lectures he had enjoyed as an undergraduate. In the following year, he proved a profound converse of

a famous theorem on the summation of series by the Norwegian mathematician Niels Henrik Abel. As he later recalled: 'On looking back this time seems to me to mark my arrival at a reasonably assured judgement and taste, the end of my "education". I soon began my 35-year collaboration with Hardy.'

THE RAMANUJAN YEARS

In early 1913, Bertrand Russell wrote a letter to a friend saying:

> In Hall I found Hardy, and Littlewood in a state of wild excitement, because they believe they have found a second Newton, a Hindu clerk in Madras on £20 a year. He wrote to Hardy telling of some results he had got, which Hardy thinks quite wonderful, especially as the man has had only an ordinary school education. Hardy has written to the Indian Office and hopes to get the man here at once.

This 'second Newton' was Srinivasa Ramanujan, who in January had written to Hardy introducing himself, explaining that he had come across Hardy's *Orders of Infinity*, and submitting eleven pages of his own mathematical discoveries for Hardy's consideration. These results, relating to prime numbers, summation of series, divergent series, and improper integrals, were supplied without proofs. Although some were imprecise or incorrect, others showed spectacular insight, and Hardy and Littlewood surmised that they must be correct, since no-one would have had the imagination to make them up. Letters were interchanged, and it soon became clear that Ramanujan was a genius of the first order, but an untutored one, ignorant of many of the most basic results in formal mathematics as taught in England.

After much further correspondence and delay, for bureaucratic and financial reasons and also because, as a devout Brahmin, Ramanujan had to overcome caste beliefs and prejudices, he set sail for England, arriving in Cambridge in April 1914. After a short while he moved to Trinity and started to work with Hardy and Littlewood, and, by June, Hardy was already communicating Ramanujan's results at a meeting of the London Mathematical Society.

At the outbreak of the First World War, Littlewood became a Second Lieutenant in the Royal Garrison Artillery and was soon seconded for ballistics work, improving the methods for calculating trajectories of anti-aircraft missiles. The mathematical intricacies were not formidable, and Littlewood's contributions were substantial. Meanwhile, he continued his own researches, collaborating on a regular basis with Hardy and producing several joint papers during the War years. He was elected a Fellow of the Royal Society in 1916.

Hardy, an ardent pacifist, stayed in Cambridge with Ramanujan, obtaining some spectacular results. One particularly fruitful area was that of partitions. The problem is to find the number $p(n)$ of ways of splitting a positive integer n into separate parts; for example, $p(4) = 5$, corresponding to the partitions $4, 3+1, 2+2, 2+1+1$ and $1+1+1+1$. Values up to $p(200) = 3\,972\,999\,029\,388$ were known, but no general formula had ever been found. Hardy and Ramanujan obtained an astonishing formula, yielding the exact value for $p(n)$ for any given value of n; it involved 24th complex roots of unity, exponentials, derivatives, and much more besides, and was based on the 'Hardy–Littlewood circle method', a powerful technique devised for tackling a range of problems in number theory. As Littlewood later commented: 'We owe the theorem to a singularly happy collaboration of two men, of quite unlike gifts, in which each contributed the best, most characteristic, and most fortunate work that was in him.' Hardy considered Ramanujan to have had the intellectual ability of an Euler or a Gauss, and later described his three-year mathematical partnership with him as the one romantic incident of his life.

Apart from his stimulating collaboration with Ramanujan, the War years were unhappy ones for Hardy. As a pacifist he was continually at odds with most of the Trinity dons, and led the protests when Bertrand Russell's lectureship was revoked by the Trinity College governing body after Russell was imprisoned for anti-War activities; Hardy later chronicled these events in his *Bertrand Russell and Trinity*.

In May 1917 Ramanujan became very ill from tuberculosis, a

result of the Cambridge climate and a poor diet. There is a well-known story of Hardy visiting Ramanujan in hospital in Putney:

> Hardy had arrived by taxi and, unable to think of what to say remarked, 'The number of my taxi-cab was 1729. It seemed to me rather a dull number.' 'No, Hardy!', replied Ramanujan, 'it is the smallest number that can be written as the sum of two cubes in two different ways' (1728 + 1 and 1000 + 729).

Through the efforts of Hardy and Littlewood, Ramanujan was elected to a Fellowship of the Royal Society and also to a Fellowship at Trinity College. Both of these honours delighted Ramanujan, but he was still very ill and returned to India in early 1919. He died on 26 April 1920, and Hardy was devastated. He wrote:

> It is difficult for me to say what I owe to Ramanujan – his originality has been a constant source of suggestion to me ever since I knew him, and his death is one of the worst blows I ever had.

In 1921 Hardy wrote Ramanujan's obituary for the Royal Society and the London Mathematical Society and in the next few years produced several papers based on, and extending, Ramanujan's contributions. He worked through Ramanujan's papers and notebooks and was one of the editors of his *Collected Papers*, published in 1927. In 1940 Hardy published *Ramanujan*, a collection of essays and lectures suggested by Ramanujan's work.

HARDY'S OXFORD YEARS

By 1919 life at Trinity College had become incongenial to Hardy. The War years, the bitter wranglings with the pro-War Trinity dons, and Council's treatment of Bertrand Russell in particular, had all taken their toll. With Ramanujan back in India and strains emerging in his partnership with Littlewood, as mentioned earlier, Hardy felt that he must get away.

In Oxford the Savilian Chair of Geometry had been frozen since William Esson's death in 1916, but was released after the War and

advertised in late 1919. Hardy applied for it and was appointed in December 1919, taking up his position in January 1920, while Littlewood succeeded Hardy as Cayley Lecturer at Cambridge. Hardy's Oxford appointment carried a stipend of £900 per annum and required him to give 42 lectures per year. Among his colleagues was the applied mathematician Augustus Love, now Sedleian Professor of Natural Philosophy, who had so inspired Hardy in Cambridge over twenty years earlier.

Although not a geometer, Hardy always gave courses in the subject, gradually adding courses on number theory, the theory of functions, and other topics. In a Presidential Address to the Mathematical Association in 1925 on 'What is geometry?', he observed:

> I do not claim to know any geometry, but I do claim to understand quite clearly what geometry is.

His only result in geometry, he once claimed, was to prove that 'if a rectangular hyperbola is a parabola, then it is also an equiangular spiral'.

Hardy's appointment was attached to New College, whose informality and friendliness enabled him to feel completely at home. Hardy led the conversation in the Common Room and everyone waited to hear what he was going to say, while he with mock seriousness organised complicated games and intelligence tests. C.P. Snow described Hardy's eleven years in Oxford as the happiest in his life, while Littlewood remarked that Hardy 'preferred the Oxford atmosphere and said they took him seriously, unlike Cambridge'. Indeed, Hardy himself claimed that, 'I was at my best at a little past forty, when I was a professor at Oxford.'

Certainly his research blossomed at this time, and he wrote a hundred papers during his eleven years in Oxford. In particular, his difficulties with Littlewood had clearly blown over, since about half of these papers were joint collaborations with him. Included in this work were eight important papers entitled *Some problems of Partitio numerorum*, inspired by Waring's problem, the topic of Hardy's Oxford

inaugural lecture in May 1920. This problem arose from an observation of Lagrange in 1770 that every positive integer can be written as the sum of four perfect squares; for example, $70 = 49 + 16 + 4 + 1$. It can similarly be proved that every positive integer can be written as the sum of nine cubes or nineteen fourth powers. The question is: Does this continue for higher powers, and how many such powers are needed?

One of the consequences of the War's savagery was that Continental mathematicians had become ostracised and Hardy felt cut off from his former colleagues across the English Channel. He wrote to the German number theorist Edmund Landau about his views on war, and Landau agreed, 'with trivial changes in sign'. Hardy boycotted international conferences whenever German or Austrian mathematicians were excluded, and he visited Germany and Scandinavia, cooperating with the peacemaking efforts of the Swedish mathematician Gösta Mittag-Leffler until the situation improved.

During these years Hardy welcomed several foreign visitors to Oxford. These included the Russian émigré Abram Besicovitch, who would eventually succeed Littlewood as Rouse Ball professor in Cambridge. Another visitor was the Hungarian mathematician Georg Pólya, who had been appointed the first international Rockefeller Fellow (on Hardy's recommendation) and who spent half the year with Hardy in Oxford and the other half with Littlewood in Cambridge; a consequence of these collaborations was the celebrated text *Inequalities* by the three of them, which appeared in 1934. As part of his continuing campaign against the Cambridge Mathematical Tripos, Hardy arranged for Pólya to sit the papers, assuming them to be so irrelevant to modern mathematics that Pólya would perform badly. Unfortunately for Hardy, Pólya's answers were so good that he would have become Senior Wrangler if he had been officially registered.

These Oxford visitors attended Hardy's weekly mathematical meetings, at which several postgraduate students cut their teeth as research mathematicians; these included Edward Titchmarsh (Hardy's successor as Savilian Professor), Mary Cartwright (later Mistress of

Girton and the first woman mathematician elected to the Royal Society), and Edward Wright (later Vice-Chancellor of the University of Aberdeen and an important collaborator of Hardy).

Meanwhile, in Cambridge, Littlewood and Harald Bohr collaborated on a monograph on number theory, but were both so exhausted when the manuscript was complete that they did not have the strength to send it to the printer and it was left for several years. By this time the subject had developed considerably, and they handed it to others to use. In 1926 Littlewood did complete a textbook, *Elements of the Theory of Real Functions*, but he never had Hardy's facility in this direction and preferred to concentrate on research papers.

The 1920s were years of recognition for Hardy and Littlewood. At the Royal Society Hardy was awarded the Royal Medal in 1920, and Littlewood received the same medal nine years later. From 1917 to 1926, Hardy became secretary of the London Mathematical Society, never missing a meeting of the Society or of its Council, and became president from 1926 to 1928; he received the Society's highest honour, the De Morgan Medal, in 1929. Hardy also became president of the Mathematical Association from 1925 to 1927, and of the National Union of Scientific Workers in 1926. In 1928 Littlewood was appointed to the newly founded Rouse Ball Chair of Mathematics in Cambridge, a very appropriate appointment, not least because he had been one of Rouse Ball's favourite pupils at Trinity. Littlewood received his first honorary degree in the same year, a Doctorate of Science from Liverpool University.

Both of them found relaxation in sporting activities. Hardy was a keen cricketer and tennis player and after leaving Oxford continued to return every summer to lead the Fellows' cricket team at New College; a well-known picture shows him leading a team of distinguished mathematicians on to the cricket field. Littlewood, an able athlete, enjoyed swimming and took up rock climbing while on holidays in Cornwall and the Isle of Wight. In 1924 he developed a taste for skiing while on vacation in Switzerland.

Hardy had a number of eccentricities. He would never wear a watch or use a fountain pen, and avoided telephones whenever possible. In a list of New Year resolutions he aimed to:

1   prove the Riemann hypothesis
2   make 211 not out in the fourth innings of the last test match at the Oval
3   find an argument for the nonexistence of God which shall convince the general public
4   be the first man at the top of Mt Everest
5   be proclaimed the first president of the USSR of Great Britain and Germany
6   murder Mussolini.

As an ardent atheist, he carried on a continual feud with God, his 'personal enemy'. He would never enter a religious building, and a clause had to be inserted into the New College by-laws allowing him to cast his vote for a new Warden without being present in the college chapel. A sheet of headed writing paper in the New College archives records the following:

> What a day. Beat you at squash (GHH 67, G 33); Good math. theorem (O.K) GHH 73, G 0 . . . Bishop dead (forgot his name) GHH 33, G 0; Ex-Chorister to be hanged GHH 61, G 10. Total score GHH 379, God 43.

In 1928–9 Hardy arranged an exchange with the American geometer Oswald Veblen, visiting Princeton University for the year and (briefly) California Institute of Technology. During this time he gave several invited lectures around the United States, while developing a liking for baseball and American football and enjoying the opportunity to work without constant interruptions.

In 1926 Hardy had successfully urged the founding of the *Journal of the London Mathematical Society*. On his return from the United States, following the death of its former editor James Glaisher, Hardy also founded a new series of the *Quarterly Journal of Mathematics*.

REUNITED AGAIN

In 1931 E. W. Hobson resigned the Sadleirian Chair in Cambridge, a post held for many years by Arthur Cayley. Although Hardy had transformed the Oxford mathematical scene, creating a school of analysis second to none, Cambridge was still the centre for mathematical research in the UK, and Hardy decided to apply. At Trinity he would be allowed to live in college rooms for the rest of his life, whereas no such facility was available in Oxford. Hardy was duly appointed, and spent the rest of his life in Cambridge with Littlewood.

In the late 1920s Hardy and Littlewood had collaborated on two highly original and influential papers, on the rearrangement of series and on a maximal theorem. These papers were to be the foundation of later research by many mathematicians. Their collaboration continued with Hardy's return to Cambridge, and from early 1932 they held weekly mathematical 'conversation classes' in Littlewood's rooms. According to Edward Titchmarsh:

> This was a model of what such a thing should be. Mathematicians of all nationalities and ages were encouraged to hold forth on their own work, and the whole exercise was conducted with a delightful informality that gave ample scope for free discussion after each paper.

During these Cambridge years Hardy and Littlewood continued to collaborate freely with other mathematicians. Hardy wrote his celebrated *An Introduction to the Theory of Numbers* (1938) with his former student Edward Wright, and his final Cambridge Mathematical Tract, *Fourier Series* (1944), with Werner Rogosinski. Meanwhile, Littlewood carried out some far-reaching work on Fourier series in an important series of papers with his brilliant student R. Paley, but this tragically came to an end with the latter's death in a skiing accident at the age of twenty-six. He also collaborated with Cyril Offord on random equations and with Mary Cartwright on some differential equations arising from electrical circuit theory. Later, during the Second World War, Cartwright and Littlewood carried out some important work on the mathematics of radar. He later wrote of this collaboration:

> Two rats fell into a can of milk. After swimming for a time one of them realised his hopeless fate and drowned. The other persisted, and at last the milk was turned to butter and he could get out.

Both Hardy and Littlewood wrote well-known books explaining the nature of mathematics to a general readership. Hardy's *A Mathematician's Apology* (1940), a personal account written at the beginning of the Second World War, is a rather sad book by a mathematician looking back as his powers are waning. In contrast, Littlewood's *A Mathematician's Miscellany* (1953) is a more joyful book, full of mathematical gems and allowing the reader to experience academic life at Trinity through his perceptive eyes.

These years continued to bring external recognition to Hardy and Littlewood. In 1939–41 Hardy again became President of the London Mathematical Society, becoming the only president to serve two terms. Littlewood succeeded him in 1941–3, having received the De Morgan Medal in 1938 and later winning the Senior Berwick Prize (1960). From the Royal Society, Littlewood received the Sylvester Medal in 1943 and the Copley Medal in 1958.

Hardy retired from the Sadleirian Chair in 1942. He continued to write, and his last book, *Divergent Series* (1948) appeared after his death. In 1947 he was elected 'associé étranger' of the Paris Academy of Sciences, one of only ten people from all nations and scientific subjects. He died on 1 December 1947, the very day that he was due to be presented with the Copley Medal by the Royal Society. His epitaph could well be this sentence from *A Mathematican's Apology*:

> I still say to myself when I am depressed, and find myself forced to listen to pompous and tiresome people, 'Well, I have done one thing you could never have done, and that is to have collaborated with both Littlewood and Ramanujan on something like equal terms.'

Littlewood retired in 1950 at the statutory age of 65. However, he was permitted to continue teaching after the mathematics faculty wrote to the General Board:

Professor Littlewood is not only exceptionally eminent, but is still at the height of his powers. The loss of his teaching would be irreparable, and it is avoidable. Permission is requested to pay a fee of the order of £100 for each term's course of lectures.

In fact, Littlewood received just £15 for each course.

For thirty years Littlewood had suffered from a nervous malady which caused him to function at a fraction of his capacity, and he spent many hours in local cinemas whiling away the tedious hours. But in 1960, a brilliant neurologist experimented with some new drugs and cured the depression. This gave Littlewood a new lease of life and at the age of 75 he started to take up lecturing invitations, particularly in the United States which he visited eight times.

Littlewood continued his mathematical researches until he was well into his eighties. He died on 6 September 1977 at the age of 92.

Further reading

Béla Bollobás (ed.). 1986. *Littlewood's Miscellany*. Cambridge: Cambridge University Press.

J.C. Burkill *et al*. 1978. John Edensor Littlewood. *Biographical Memoirs of Fellows of the Royal Society*, **24**, 323–67.

G.H. Hardy. 1966–79. *Collected Papers* (7 vols.). Oxford: Oxford University Press.

Robert Kanigel. 1991. *The Man Who Knew Infinity*. New York: Charles Scribner.

J.E. Littlewood. 1982. *Collected Papers* (2 vols.). Oxford: Oxford University Press.

E.C. Titchmarsh. 1949. Godfrey Harold Hardy, *Obituary Notices of Fellows of the Royal Society*, **6**, 447–70.

Godfrey Harold Hardy (1877–1947), British mathematician: developed new work in analysis and number theory.

Hardy was the son of an art teacher; he was a precocious child, whose tricks included factorising hymn-numbers during sermons. His early mathematical ability won him a scholarship to Trinity College, Cambridge, where he was elected a Fellow. In 1920 he became Savilian Professor of Geometry at Oxford, but returned to Cambridge in 1931 as Sadleirian Professor of Pure Mathematics.

Hardy was an excellent teacher and introduced a modern rigorous approach to analysis. He encouraged the young Indian genius Srinivasa Ramanujan (1887–1920), bringing him to Cambridge to do research. Hardy was a staunch

anti-Christian, a firm friend of Bertrand Russell, and a passionate and talented cricketer and real tennis player.

John Edensor Littlewood (1885–1977) lectured at the University of Manchester from 1907 to 1910. He became a Fellow of Trinity College, Cambridge (1908), returning there in 1910. For thirty-five years he collaborated with Hardy working on the theory of series, the Riemann zeta function, inequalities and the theory of functions. The collaboration led to a series of papers using the Hardy–Littlewood–Ramanujan analytical method.

Littlewood was elected a Fellow of the Royal Society in 1916. He received the Royal Medal of the Society in 1929, and in 1943 received the Sylvester Medal of the Society in recognition of his mathematical discoveries and supreme insight in the analytic theory of numbers. He also received the Copley Medal of the Society in 1958 in recognition of his distinguished contributions to many branches of analysis, including Tauberian theory, the Riemann zeta-function, and non-linear differential equations.

Robin Wilson is a Senior Lecturer in Mathematics at the Open University, a Fellow of Keble College, Oxford, and a Visiting Professor in the History of Mathematics at Gresham College, London. He has written and edited a number of books on graph theory and combinatorics, including *Introduction to Graph Theory* (Longman, 1996) and *An Atlas of Graphs* (Oxford, 1998), and is currently preparing a book on the history of combinatorics. He is also very involved with research into the history of British mathematics (especially of the Victorian era), and is a co-editor of *Let Newton be!* (Oxford, 1988) and *Oxford Figures* (Oxford, 2000).

15 Arthur Stanley Eddington
MALCOLM LONGAIR

INTRODUCTION

The most distinguished astrophysicist of his time

(S. Chandrasekhar 1983)

It is difficult to understand why Eddington, who was one of the early enthusiasts and staunchest advocates of general relativity, should have found the conclusion that black holes may be formed during the course of the evolution of stars so unacceptable. But the fact is that Eddington's supreme authority in those years effectively delayed the development of fruitful ideas along these lines for some thirty years.

(S. Chandrasekhar 1979)

No one who knew Eddington would . . . remotely imagine his being unpleasant. The last few years have seen most unfortunate, completely misleading accounts of the matter.

(W.H. McCrea 1993)

May I conclude by assuring Prof. Eddington it would give me great pleasure if he could remove a long-standing source of friction between us by abstaining in future from making wild attacks on my work which he cannot substantiate, and by making the usual acknowledgments whenever he finds that my previous work is of use to him? I attach all the more importance to the second part of the request, because I find that some of the most fruitful ideas which I have introduced into astronomical physics – e.g, the annihilation of matter as a source of stellar energy, and the highly dissociated atoms

ARTHUR STANLEY EDDINGTON 221

FIGURE 15.1 Sir Arthur Stanley Eddington (1882–1944), 1932.
Source: Photograph courtesy of The Bettmann Archive, UPI/Bettmann Newsphotos.

and free electrons as the substance of the stars – are by now fairly generally attributed to Prof. Eddington.

(J.H. Jeans November 1926)

He was so convinced, as early as 1932 when he wrote this small book (*The Expanding Universe*, 1933) that he evidently considered himself justified in giving the impression that he had fully established the results, and that here he is merely suggesting why they should strike his reader as wholly plausible.

Frankly, it is a maddening production. The reader can never be quite sure when he is being invited to follow a serious argument, and when he is being – oh so delicately – conned.'

(W.H. McCrea 1987).

Eddington's imagination and receptivity was formed along paths that valued search and unknowing at once, that distrusted fixed images, and whose own presiding metaphor was that of the "inward light".

(G. Beer 1998).

In the age of reason, faith yet remains supreme: for reason is one of the articles of faith.

(A.S. Eddington 1939).

Sir Arthur Stanley Eddington is beyond question one of the great astrophysicists of the twentieth-century. Indeed, when Sir William McCrea heard of the title of Chandrasekhar's memoir of Eddington, *The Most Distinguished Astrophysicist of his Time* (1983), he wondered who else at any time had made more distinguished contributions to the subject. And yet, the above quotations indicate the very wide range of feelings and opinions aroused by Eddington as a scientist, as a populariser of science, and as an individual. The title of David S. Evans' biography *The Eddington Enigma* (1998) sums up the case. What is unquestionably the case is that Eddington made astrophysics an exact

science and showed how extreme conditions, not available in the laboratory, could be explored in astrophysical environments.

A BRIEF CURRICULUM VITAE

Eddington was born in 1882 at Kendal on the border of the Lake District. His father died when he was only two and he and his sister Winifred were brought up by their mother at Weston-super-Mare where the family settled. The budding don and popular lecturer revealed himself early as illustrated by the charming story recounted by Eddington's biographer A. Vibert Douglas

> as a small boy he would announce a lecture on some such topic as the Moon or Jupiter, pinning a notice at the foot of the attic stairs, up which at the appointed hour a faithful and devoted old servant would climb in order that there might be at very least an audience of one!

Fortune smiled on Eddington throughout his school and University education. At Brynmelyn school, he was taught by gifted teachers in mathematics, physics, natural history, history, and literature. At the age of sixteen, he won an entrance scholarship to Owen's College, now the University of Manchester, where he was taught physics by Arthur Schuster and mathematics by Horace Lamb. In 1902, he won an entrance scholarship to read mathematics at Trinity College, Cambridge. He attended the lectures of the celebrated mathematicians E.T. Whittaker, A.N. Whitehead, and E.W. Barnes, as well as being trained in the techniques for success in the Mathematical Tripos by the most distinguished mathematical coach of his time, R.A. Herman. In only his second year, he became Senior Wrangler, the first time a second-year student had attained that distinction.

In 1906, he took up the appointment of chief assistant at the Royal Greenwich Observatory, where he obtained a thorough training in practical astronomy and began his theoretical studies in stellar dynamics. In 1913, he returned to Cambridge as Plumian Professor of Astronomy and then Director of the Observatories – he was to remain

in Cambridge for the rest of his career. His mother and sister joined him in the Director's residence in the main Observatory building. He never married. He became a public celebrity following the success of the 1919 eclipse expedition to verify the general theory of relativity and became a famed public speaker and author of popular expositions of physics and cosmology. The brilliant rhetoric of his public lectures and books, contrasted with the reputed dullness and hesitancy of his university lectures. He was knighted in 1930 and awarded the Order of Merit in 1938. He died in 1944 following a major operation from which he did not recover.

His mother came from a Quaker family and he remained a committed Quaker throughout his life. His Quaker upbringing was to have a profound influence upon his thinking, particularly in the latter part of his career when he published books discussing the philosophical implications of studies of fundamental physics and cosmology. His Quaker beliefs also meant that he could not take human life and, during the First World War, he adopted the position of a conscientious objector. This required considerable moral courage, particularly after 1917 when conscription was enforced and a number of his colleagues and staff were called up, some of them never to return. The intervention of the Astronomer Royal, Sir Frank Dyson, persuaded the authorities to allow Eddington a twelve-month deferment to work on preparations for the 1919 eclipse expedition. As a result, Eddington was allowed to continue with his astronomical and astrophysical researches at a time when he was at the very peak of scientific creativity.

ASTROPHYSICS IN THE FIRST DECADES OF THE TWENTIETH-CENTURY

When Eddington was born, astrophysics scarcely existed. Astronomy consisted primarily of precise positional astronomy supplemented by visual estimates of the brightnesses of the stars. In this role, astronomy was of practical importance for precise time keeping and for the preparation of nautical almanacs for navigational purposes. The birth of astrophysics is traditionally dated to 1860 and the publication of

Robert Bunsen and Gustav Kirchhoff's determination of the chemical composition of the Sun by spectroscopic analysis. By the 1890s, the optical spectra of large numbers of stars could be recorded photographically, the most heroic efforts being those undertaken by Edward Pickering and his band of lady assistants at the Harvard College Observatory. Their efforts marked the beginning of the systematic analysis of the spectra of very large numbers of stars and the development of schemes for putting some order into the diversity of features exhibited by them. The *Astrophysical Journal*, nowadays the pre-eminent international journal of astronomy and astrophysics, was founded in 1895 by George Ellery Hale and James E. Keeler, specifically dedicated to the new science of astrophysics, meaning the study of the physics of the Sun, planets, stars and nebulae by means of spectrographic observations with optical telescopes.

A striking feature of astronomy in the early years of the twentieth-century was just how few astronomers, let alone astrophysicists, there were. The International Astronomical Union was founded in 1919 and, at the first General Assembly held in Rome in 1922, there were just over 200 members from nineteen adhering countries. Of these members thirty were from the UK and only a handful of these would be considered astrophysicists, rather than astronomers. By 1938, the total numbers had risen to 550 from twenty-six countries. Thus, throughout Eddington's career, astrophysics remained the province of a remarkably small number of professional astronomers.

Even by 1907, when Eddington took up his post at the Royal Greenwich Observatory, understanding of the physics of the stars was in a remarkably primitive state. According to the picture promoted by Hermann von Helmholtz and William Thomson (later Lord Kelvin) in the 1850s and 1860s, the gradual contraction of the Sun could provide an enormous reservoir of gravitational potential energy. The Sun was conceived of as a convective mass of liquid which gradually contracted and cooled. If gravitational contraction was indeed the source of energy for the Sun, its age could be estimated and, in the 1880s, Thomson found an age of twenty million years. This value was considerably

shorter than the estimates of the age of the Earth from stratigraphic analyses, although Thomson argued that these were of dubious reliability. What were to prove to be the first reliable estimates of the age of the Earth were made in 1904 by Ernest Rutherford, using the relative abundances of radioactive and stable isotopes of the heavy elements. He found that some radioactive minerals must be between about 500 and 1000 million years old. In the 1880s, Norman Lockyer, the founder of *Nature*, suggested that the different stellar types formed a temperature sequence, and made the first hesitant steps towards placing the stars in an evolutionary sequence, but it was largely guesswork.

In the 1860s, J. Homer Lane was the first to treat the internal structure of the Sun as a gaseous body, but he could not reproduce its observed surface properties. In the late 1870s, similar calculations were carried out independently by Augustus Ritter at Aachen who identified the initial phase of evolution of the star as the contraction of a gaseous sphere, which subsequently cooled according to the Kelvin–Helmholtz prescription. The culmination of these physical models was the treatise *Gaskugeln* by Robert Emden, published in 1907. The intention of the treatise was to attract students at the Technische Hochschule in Munich into theoretical physics by giving as 'practical examples' the internal structure of the stars!

Eddington had the great fortune of being the right person, in the right place at the right time. His combination of deep physical insight and intuition and his remarkable mathematical ability, combined with his experience of observation at the Royal Greenwich Observatory and at the Cambridge Observatories, were ideally matched to the revolution which was about to overtake astronomy. According to Henry Norris Russell, whose theory of stellar evolution was comprehensively demolished by Eddington,

> Several investigators – Jeans, Kramers, Eggert – have contributed to this field, but much the largest share is Eddington's.

Between 1916 and 1924, he published over a dozen papers which were collected and extended in his great book of 1926, *The Internal*

Constitution of the Stars. These deep insights are the basis of the modern theory of stellar structure and evolution. The simplest way of summarising Eddington's achievement is to paraphrase Chandrasekhar's assessment.

1. The treatment of stars as gas spheres had assumed that the internal pressure was due to the pressure of the hot gas. Eddington showed that, although this might be the case for low-mass stars, the pressure of the radiation itself becomes more and more important as the mass of the star increases. In the limit of the most massive stars known, radiation pressure is dominant and ultimately leads to an upper limit to their masses.

2. In those regions of a star in which radiation is the means of transporting energy, rather than convection, the distribution of energy sources and the opacity of the stellar material determine the temperature gradient inside the star which enables its central temperature to be estimated.

3. The term, opacity of stellar material, means the difficulty radiation has in escaping from the star because of absorption and scattering. Eddington realised that the most important source of opacity was the ionisation of the innermost shells of highly ionised atoms.

4. Even if one adopts the minimum possible value of the opacity, that associated with the scattering of radiation by free electrons, there is an upper limit to the luminosity which a star of a given mass can have. Nowadays, this maximum luminosity is referred to as the *Eddington limit* and it plays a central role in many different aspects of astrophysics.

5. In normal stars like the Sun, the relation between a star's luminosity, its radius and its surface temperature is not very sensitive to the distribution of energy sources within the star. Therefore, a relation between these quantities can be found which can be compared with observation, even in the absence of a detailed knowledge of the energy sources of the stars.

6   The burning of hydrogen into helium is the most likely source of stellar energy.

These deep insights were not gained without a struggle and many of these issues were the subject of heated debate between Eddington, James Jeans, and E.A. Milne. In his first paper on the internal structure of the stars, Eddington assumed that the mean atomic mass of the particles was 54, meaning that the star was composed of iron. This was quickly corrected by Jeans who pointed out that, at the high temperature of the stellar interior, 'a rather extreme state of disintegration is possible'. In Eddington's next paper, a mean atomic weight of 2 was adopted, corresponding to the complete ionisation of the atoms, assuming that there is no hydrogen present.

In 1917, Eddington still adhered to the prevailing picture that stars like the Sun are liquid spheres, but his theory could be applied to the gaseous envelopes of the giant stars which had been identified by Ejner Hertzsprung and Henry Norris Russell. In his paper of that year, Eddington showed that, if the release of gravitational energy was the source of the luminosity of giant stars, they could not radiate for more than 100,000 years, very much less than the age of the Earth. The opportunity of testing Eddington's theory of red giants arose in 1919. Albert A. Michelson had been developing the techniques of optical interferometry for almost thirty years and George Ellery Hale, director of the Mount Wilson Observatories, decided that the 100-inch Hooker telescope should be equipped with a Michelson stellar interferometer to determine the separations of close binary stars, if not the diameters of the stars themselves. Michelson did not know how long the interferometer should be but, being aware that the instrument was in the process of construction, Eddington used his theory of the structure of red giants to predict the angular size of the bright giant star Betelgeuse. In the light of this prediction, Michelson built a 6 metre interferometer which was mounted on the top ring of the 100 inch telescope. On the night of 13 December 1919, Francis G. Pease and J.A. Anderson measured the angular diameter of Betelgeuse to be 0.047 arcsec, just slightly

less than Eddington's prediction. This observation confirmed the large diameters of red giant stars.

In 1919, Eddington discovered to his surprise that he could account for the observed relation between mass and luminosity for stars similar to the Sun, if he applied his theory of stellar structure to these stars as well. The implications were profound – stars like the Sun could not be incompressible liquid spheres, but rather gaseous spheres, cutting the foundation from under the standard picture. This conclusion was vigorously opposed by Jeans who argued that the result was spurious since it ignored the process of energy generation inside the stars. Fortunately, it turns out that the mass-luminosity relation is remarkably independent of the precise process of energy production.

The origin of the energy for a star like the Sun was a matter of controversy. Jeans proposed that the source of energy in the Sun was radioactive decay. In his early papers, Eddington advocated the annihilation of matter as an inexhaustible source of energy for the stars. In 1920, he found a better energy source, as described in a remarkably prescient paragraph of his Presidential Address to the Mathematical and Physics Section of the British Association for the Advancement of Science at its Annual Meeting, which was held in Cardiff.

> F.W. Aston's experiments (at the Cavendish Laboratory) seem to leave no room for doubt that all the elements are constituted out of hydrogen atoms bound together with negative electrons. The nucleus of the helium atom, for example, consists of 4 hydrogen atoms bound with two electrons. But Aston has further shown conclusively that the mass of the helium atom is less than the sum of the masses of the 4 hydrogen atoms which enter into it; ... There is a loss of mass in the synthesis amounting to about 1 part in 120, the atomic weight of hydrogen being 1.008 and that of helium 4. ... Now mass cannot be annihilated, and the deficit can only represent the mass of the electrical energy set free in the transmutation. We can therefore at once calculate the quantity of energy liberated when helium is made out of hydrogen. If 5 per cent of the star's mass consists initially

of hydrogen atoms, which are gradually being combined to form more complex elements, the total heat liberated will more than suffice for our demands, and we need look no further for the source of a star's energy.

It was not appreciated then that most of the mass of normal stars is in the form of hydrogen and, despite the fact that the evidence for this became overwhelming by the 1930s, Eddington was reluctant to accept the observational evidence.

RELATIVITY AND WHITE DWARFS

Einstein's discovery of the general theory of relativity, the relativistic theory of gravity, was communicated to the Berlin Academy of Sciences in 1915. Because of the War, direct communication with physicists in Germany was not possible, but the papers were forwarded to Eddington by Willem de Sitter, a personal friend of Eddington's in neutral Holland. The theory is of considerable mathematical complexity, but as Einstein stated in the last paragraph of his final paper on the theory, 'scarcely anyone who has fully understood this theory can escape from its magic'. Eddington was the ideal expositor of these ideas in English and within two years had written his *Report on the Relativity Theory of Gravitation* for the Physical Society of London. This is a brilliant exposition of Einstein's theory, which made three predictions which contrasted with the expectations of Newtonian gravity. First, the elliptical orbit of the planet Mercury should precess by 43 arcsec per century because of the bending of space-time about the Sun, a value in excellent agreement with observed estimates of the precession, which dated from Le Verrier's analysis of 1859. Second, the theory predicted the deflection of light by massive bodies because of the curvature of space-time in their vicinity. The predicted deflection of light rays for stars just grazing the limb of the Sun amounted to 1.75 arcsec; according to 'Newtonian' theory, the deflection would have amounted to only half this value. Third, the light escaping from massive bodies is shifted to slightly

longer wavelengths, the phenomenon known as the *gravitational redshift*.

Einstein's prediction of a deflection of 1.75 arcsec provided the stimulus for the famous eclipse expeditions of 1919 led by Eddington and Crommelin, one to Sobral in Northern Brazil and the other to the island of Principe, off the coast of West Africa. This was a unique opportunity for carrying out this test of the theory since the eclipse took place when the alignment of the Moon and the Sun was in the direction of the Hyades star cluster and so there were many suitable bright stars which could be used to estimate the deflection. The results were in good agreement with Einstein's prediction, the Sobral result being 1.98 ± 0.12 arcsec and the Principe result 1.60 ± 0.3 arcsec, both significantly different from the Newtonian prediction of 0.83 arcsec. Because of the technical difficulty of these observations, the precise value of the deflection remained a somewhat controversial issue. Nonetheless, in the post-War atmosphere, this remarkable triumph for general relativity raised both Einstein and Eddington to positions of public prominence, and from 1919 onwards Eddington's authority as an astrophysicist, cosmologist, and theorist was unchallenged in the public arena.

The third test of general relativity became possible with the discovery of white dwarf stars. In 1910, Henry Norris Russell, Pickering, and Williamina Fleming discovered that the faint companion of o-Eridani was a very low-luminosity star, but had the type of spectrum which was only associated with the hottest stars then known. Walter Adams discovered another example in 1915, the faint companion of Sirius A known as Sirius B. Eddington realised that these observations implied that white dwarf stars had to be very dense indeed, his estimates of their mass density being about 100 million kg m^{-3}. Eddington argued that there was nothing inherently implausible about such large densities. Matter at high temperatures inside stars would be completely ionised and so there was no reason at that time why the matter could not be compressed to much higher densities than typical terrestrial densities. In his paper of 1924, he estimated the gravitational redshift which would be expected from such a compact

star according to general relativity and found that it corresponded to a Doppler shift of the spectral lines to longer wavelengths of 20 km s^{-1}. Adams made very careful spectroscopic observations of Sirius B with the 100-inch telescope in 1925 and a gravitational redshift of 19 km s^{-1} was measured. Eddington was jubilant:

> Prof. Adams has thus killed two birds with one stone. He has carried out a new test of Einstein's theory of general relativity, and he has shown that matter at least 2000 times denser than platinum is not only possible, but actually exists in the stellar Universe.

The theory of white dwarfs was one of the first triumphs of the new quantum theory of statistical mechanics as applied to astrophysics. In 1926, Ralph Fowler used these concepts to derive the equation of state of a cold degenerate electron gas which provides the pressure support for low-mass white dwarfs. In 1929, Wilhelm Anderson showed that the degenerate electrons in the centres of white dwarfs with mass roughly that of the Sun become relativistic. Anderson and Edmund Stoner realised that the consequence was that there do not exist equilibrium configurations for degenerate stars with mass greater than about the mass of the Sun.

The most famous analysis of this problem was carried out by S. Chandrasekhar who began working on it before he arrived to take up his fellowship at Trinity College, Cambridge in 1930. He derived the key result while on board the ship *Lloyd Triestino* which was taking him as a nineteen year-old from Bombay to London. He found the crucial result that there is an upper limit to the mass of stable white dwarfs of about one and a half times the mass of the Sun. This mass is known as the *Chandrasekhar mass* – there is nothing to prevent more massive degenerate stars from collapsing to a state of complete gravitational collapse. This conclusion was vigorously challenged by Eddington and led to the famous dispute with Chandrasekhar. Eddington found the idea of complete gravitational collapse unacceptable and repudiated Chandrasekhar's result in public at the meeting of the Royal Astronomical Society on 11 January 1935.

> Dr Chandrasekhar has got this result before, but he has rubbed it in in his last paper; and, when discussing it with him, I felt driven to the conclusion that this was almost a *reductio ad absurdum* of the relativistic degeneracy formula. Various accidents may intervene to save the star, but I want more protection than that. I think there should be a law of Nature to prevent a star behaving in this absurd way!

It is clear from the discussion that Eddington realised that what we would now call a black hole was the natural outcome of gravitational collapse, but he objected instinctively to what he called elsewhere 'this stellar buffoonery'. Chandrasekhar took the rebuff badly and it rankled for many years, despite his lasting respect for and friendship with the older man.

COSMOLOGY AND THE FUNDAMENTAL THEORY

Einstein's discovery of general relativity enabled him to derive the first fully self-consistent model for the Universe as a whole. In 1917, long before it was known that the distribution of galaxies is expanding apart, Einstein found a solution of the equations of general relativity corresponding to a closed, finite static Universe, provided a repulsive cosmological term, known as the Λ-term, was introduced into the equations to counteract the attractive force of gravity. Einstein believed that, by introducing this term, he had incorporated *Mach's Principle*, according to which the local reference frame is defined by the distribution of matter on the large scale in the Universe, into general relativity. It was soon shown by de Sitter that this conclusion was incorrect, since there existed solutions of Einstein's equations, even if there were no matter present in the world model. In 1922–4, Aleksander Friedman showed that Einstein's equations had perfectly satisfactory unstable solutions in the sense that the whole Universe could be either expanding or contracting. These results were discovered independently by Georges Lemaître in 1927. Hubble's discovery of the expansion of the system of galaxies in 1929 provided convincing

evidence that these unstable solutions were consistent with the observed large-scale kinematics of the Universe. With this discovery, Einstein retracted his introduction of the cosmological term.

This was, however, very far from the end of the story because there remained one grave problem for the models in which the cosmological constant Λ is set equal to zero. If $\Lambda = 0$, the age of the Universe must be less than the inverse of Hubble's constant, the constant in the velocity–distance relation. Using Hubble's data, the age of the Universe had to be less than two billion years old, a figure in conflict with the age of the Earth derived from the ratios of abundances of long-lived radioactive species, which suggested a significantly greater age.

Eddington and Lemaître recognised that this problem could be eliminated if the cosmological constant is positive. The effect of a positive cosmological constant is to counteract the attractive force of gravity when the Universe grows to a large enough size. Among the solutions, there are cases corresponding to stationary Einstein Universes in the past rather than at the present day. According to these models, the Universe could remain in the static Einstein state for an arbitrarily long period in the past and then begin to expand away from that state under the influence of the cosmological term. In this type of *Eddington–Lemaître* model, the age of the Universe could be arbitrarily long. As Eddington expressed it, the Universe would have a 'logarithmic eternity' to fall back on and so resolve the conflict between estimates of Hubble's constant and the age of the Earth. The initial singularity present in the expanding models without the cosmological term was abolished, a clear attraction for Eddington. As he wrote in *New Pathways in Science*:

> Philosophically the notion of an abrupt beginning of the present order of Nature is repugnant to me, as I think it must be to most.

For roughly the last two decades of his life, these ideas became inextricably tied up with his ideas about the fundamental laws of physics. Quantum mechanics had only arrived on the scene in 1925–7, thanks to the efforts of Schrödinger, Heisenberg, and their colleagues.

Built into this new scheme of things was Heisenberg's Uncertainty Principle, which set fundamental limits to the precision with which certain pairs of physical quantities could be measured simultaneously. In 1928, Dirac, working in Cambridge, published his theory of the electron with the remarkable predictions of its spin, its magnetic moment, and the existence of its antiparticle, the positron, which was discovered in 1932 in the experiments carried out by Carl Anderson. Also in 1932, Chadwick discovered the neutron and identified it with a constituent of nucleus, along with the proton.

Eddington became deeply involved in the study of Dirac's theory of the electron and the types of mathematics needed to describe its remarkable spin properties. He developed his own version of the theory, involving a class of numbers which he called E-numbers – these formed a particular example of what are known as Clifford algebras. One of the key motivations of Eddington's approach to the theory was to account for the value of the dimensionless number known as the *fine-structure constant*, which appears throughout the quantum theory of the electron and has the value $\alpha = e^2/\hbar c$, where $\hbar$ is Planck's constant divided by 2π, c is the speed of light, and e is the charge of the electron in cgs units. In the early 1930s, the inverse of the fine-structure constant had a value close to 136, a whole number of special significance for Eddington's theory of the electron. As noted by David Evans, '136 is the number of independent elements in a 16×16 symmetric matrix and represented the number of degrees of freedom of a two electron system'.

To this was added the idea that there are some very large dimensionless numbers which appear in physics and cosmology. They are so large and so similar in value that Eddington argued that they must be related. Thus, the square of the ratio of strengths of the electrostatic and gravitational forces is more or less the same as the number of particles in an Eddington–Lemaître Universe, about 10^{80}, which in turn is directly related to the value of the cosmological constant Λ. The latter is uniquely defined in an Eddington–Lemaître world model once the value of the present rate of expansion of the Universe is known. Thus, for Eddington, the cosmological constant played a central role in his

vision of a physical theory unifying the very large and the very small. In his popular book *The Expanding Universe*, based upon his popular radio lecture, he stated

> To drop the cosm(olog)ical constant would drop the bottom out of space.

Eddington never deviated from his belief that this line of reasoning held within it a unified picture for processes on the scale of protons and electrons and those on the scale of the Universe itself. The *Fundamental Theory* represented his bold attempt to formulate a new type of mathematics which would encompass everything. As experimental data showed that the fine structure was closer to 1/137 and then to the value 1/137.03, Eddington produced ever more complicated refinements of the theory, but the appeal of the uniqueness of the theory was lost as more or less arbitrary 'correction factors' were added to a theory which was already mathematically complex and physically obscure. To many, his approach smacked of numerology and few thought it worthwhile to pursue this line of reasoning seriously.

For his fellow scientists, the problems must have been compounded by his philosophical writings in which he asserted that

> Generalisations that can be reached epistemologically have a security which is denied to those that can only be reached empirically.

Such a view can scarcely have appealed to a generation of physicists who were coming to terms with the completely new set of concepts of relativistic quantum mechanics to account for the experimentally determined properties of matter at the atomic level and their strange quantum behaviour. In 1937, Eddington was among those condemned, along with Milne and Dirac, by Herbert Dingle in his passionate plea against what he called 'Modern Aristotelianism'. There can be no doubting, however, the sincerity of Eddington's striving to reconcile his lifelong attachment to the Quaker world-view and his vision of a comprehensive fundamental theory which would ultimately be epistemologically based.

RESOLVING THE ENIGMAS

The conflicting visions of Eddington, with which this story began, begin to make more sense in the light of his intellectual history. His extraordinary powers as a mathematical physicist enabled him to make some of the greatest contributions to modern astrophysics. The clarity of his vision in isolating the key features needed to understand the physics of the stars is scientific virtuosity of highest calibre and will forever remain a stunning monument to the power of pure reason in opening up entirely new vistas and expanding the domain of physical enquiry.

It is striking, however, how much of his work was done entirely on his own without the benefit of collaborators. One may surmise that his training to become Senior Wrangler bred in him a self-confidence in his abilities to tackle any problem and a degree of intellectual self-assurance which could easily be misinterpreted. The ferocious interchanges between Eddington, Jeans, and Milne, faithfully recorded in the pages of *The Observatory*, make startling reading, even by comparison with today's outspoken standards. But this was simply the rhetoric of deeply involved theorists, who were members of a rather small club of theoretical astrophysicists. My own experience is that the meetings of the Royal Astronomical Society maintain a tradition of vigorous debate which should not to be taken personally. It would be wrong to believe that discussion in Cambridge today is any less robust. But it must have been a searing experience for the nineteen year-old Chandrasekhar coming from a high-caste family background in India to be confronted with the sharpness and barbed wit of Cambridge scientific discourse. It was not until long after Eddington's death that Chandrasekhar returned to general relativity and wrote his monumental *Mathematical Theory of Black Holes*.

I had the opportunity of discussing Cambridge astronomy of the 1930s with Sir William McCrea and Lyman Spitzer in the years just before they died. Both attested to Eddington's charm and his lack of malice. Both agreed that most of those who listened to the debate were on Chandrasekhar's side and that Eddington by the 1930s was becoming more and more in a minority of one in many of his views.

How is one to reconcile the shy and almost inarticulate university lecturer with the public image of a brilliant speaker, a scientific mind at its sharpest and most provocative, and with his deeply held Quaker world view? The quotation by Gillian Beer cited at the head of this essay might suggest a dichotomy between the rigour of his thought as a mathematical physicist and his ultimately unsuccessful attempts to reconcile theoretical physics with his deeply personal Quaker beliefs. To those of us working in the field, the contradictions are perhaps not so startling. Eddington's brilliant public performances were the result of careful preparation of a finely honed text and contrasted with the reported dullness of his University lectures. The passion, emotion and individuality of approach are the common currency of astronomical research and what enlivens the subject, remarks that apply equally to any of the frontier sciences. As my colleague Douglas Gough tells his students,

> Never approach any problem with an unbiased mind.

Eddington is a supreme example of this typically Cambridge approach, which can lead to the ultimate in scientific insight and creativity, and also to blind alleys which most will seek to avoid. We celebrate the former and do not allow the latter to tarnish its glory.

Further reading

G. Beer. 1998. Eddington and the Idiom of Modernism. In *Science, Reason and Rhetoric*, ed. H. Krips, J.E. McGuire and T. Melia, p.295. Pittsburg, PE: University of Pittsburg Press.

S. Chandrasekhar. 1983. *Eddington: The Most Distinguished Astrophysicist of his Time*. Cambridge: Cambridge University Press.

D.S. Evans. 1998. *The Eddington Enigma*. Princeton, NJ: Xlibris Corporation.

C.W. Kilmister. 1966. *Sir Arthur Eddington*. Oxford: Pergamon Press.

M.S. Longair. 1995. Astrophysics and Cosmology. In *20th Century Physics*, eds. L.M. Brown, A. Pais and A.B. Pippard, Vol. 3, 1691–1821. Bristol, Philadelphia and New York: Institute of Physics Publishing and the American Institute of Physics Press.

A. Vibert Douglas. 1956. *Arthur Stanley Eddington*. London: Thomas Nelson & Sons.

Arthur Stanley Eddington (1882 – 1944) an astrophysicist, was an outstanding student first at Manchester then at Cambridge, where he later became Director of the Observatory. The internal structure of stars was an area of study pioneered by Eddington. He realised that there is an upper limit on the mass of a star, above which the balance between gravitation and radiation pressure could not be maintained. He also explained the mass-luminosity relationship for stars, and provided some of the most important observational evidence for Einstein's theory of relativity.

Malcolm Longair is Jacksonian Professor of Natural Philosophy and Head of the Cavendish Laboratory, University of Cambridge. From 1980 to 1990, he held the joint posts of Astronomer Royal for Scotland, Regius Professor of Astronomy of the University of Edinburgh and Director of the Royal Observatory, Edinburgh. He is a Professorial Fellow and Vice-President of Clare Hall, Cambridge. He was President of the Royal Astronomical Society 1996–8. His primary research interests are in the fields of high-energy astrophysics and astrophysical cosmology.

16 Paul Dirac: A quantum genius
HELGE KRAGH

Paul Dirac, the distinguished theoretical physicist and Nobel laureate of 1933, was for most of his active life closely related to Cambridge University. When he retired from his position as Lucasian Professor in 1969, he had been at the university for forty-six years. He then moved to Florida, but frequently returned to St John's college for visits. Although he travelled widely and often stayed at foreign universities, his home base was always Cambridge University. Yet Dirac was first of all his own, not a 'Cambridge man', and his great scientific accomplishments were only loosely connected with the Cambridge environment and his position at the university. His colleague Nevill Mott once remarked that, 'He [Dirac] is one of the very few scientists who could work even on a lonely island if he had a library and could perhaps even do without books and journals.'

Dirac's scientific contributions covered several fields of theoretical physics, including cosmology and the theory of general relativity, but he is best known for his pioneering works in quantum mechanics and quantum electrodynamics. He made most of his remarkable discoveries as a young man, between 1925 and 1934, after which period of amazing creativity he increasingly moved away from mainstream physics. Had he died thirty-two years old, he would still be remembered as one of the greatest physicists ever, comparable to giants such as Newton, Maxwell, and Einstein.

THE DISCOVERY OF A NEW PHYSICS
Young Paul Dirac – or Paul Adrien Maurice Dirac, to be precise – seemed not destined to become an outstanding innovator of physics. Born on 8 August 1902, from 1918 to 1921 he studied electrical engineering at Bristol University, but after graduation he was unable to find

FIGURE 16.1 Paul Adrien Maurice Dirac.
Source: AIP Emilio Segrè Visual Archives, gift of Mrs Mark Zemansky (2/1986).

a job as an engineer. He was awarded a minor scholarship by St John's college, but was forced to decline it because it did not cover the expenses of studying in Cambridge. After further studies at Bristol University, now in applied mathematics, in the fall of 1923 he entered Cambridge University as a research student, funded by a grant from the Department of Scientific and Industrial Research (DSIR). He was admitted to St John's College and assigned Ralph Fowler as his supervisor.

At that time Dirac had a thorough knowledge of electromagnetism and the theory of relativity, but knew very little about the new quantum theory of the atom that Niels Bohr had pioneered a decade earlier. Until that time, he recalled, 'the atoms were always considered as very hypothetical things by me'. Fowler was one of the few British physicists who mastered the quantum theory of atomic structure, a

branch of theoretical physics dominated by physicists from Germany and Scandinavia. Partly from Fowler and partly from own studies of textbooks and research papers, Dirac quickly caught up with the new subject. He found it fascinating and within a year he had published his first papers on quantum and atomic theory.

Thanks to the stimulating Cambridge environment, Dirac widened his scientific perspective and came to meet people with interests similar to his own. There were several opportunities to join social or academic clubs, but the shy Dirac restricted his attendance to two of the physics clubs, known as the $\nabla^2 V$ Club and the Kapitza Club. Basically an introvert, he deliberately kept away from extra-mural activities that might disturb the studies on which he concentrated singlemindedly. 'I concentrated all my energy in trying to get a better understanding of the problems facing physicists', he recalled. 'I was not interested at all in politics . . . [and] confined myself entirely to the scientific work, and continued at it pretty well day after day.' His quiet and secluded life was in harmony with the environment and traditions of Cambridge theoretical physics. The few students usually sat alone in their college rooms or in the library, and were not expected to have social or professional contacts.

Dirac and the American physicists John Slater, then a postdoctoral research student, followed some of the same courses; characteristically, without talking together or knowing the existence of one another. Only years later did they realise that they had followed the same course. Mott, three years younger than Dirac, wrote that 'for students there was nowhere to sit, except in the rather small and squalid library . . . the tradition was that we should sit in our college rooms and think'. For Mott, it was 'a terribly isolated business' to be a student at Cambridge. Dirac did not find the isolation terrible at all. He had some contact with Cambridge physicists, such as Fowler, Ebenezer Cunningham, and Edward Milne, but almost none with other students; and the few contacts he had did not evolve into friendships. Dirac was a loner.

In the late summer of 1925, Dirac became acquainted with

Werner Heisenberg's new theory of atomic structure. Although he recognised Heisenberg's work to be of the greatest importance, he also found it unclear and complicated, and therefore decided to develop his own formulation of quantum mechanics. He had his paper on 'The fundamental equations of quantum mechanics' ready in early November. It was followed by a series of other papers in which Dirac formulated his theory in still more general and abstract versions, and also applied it to specific problems of physics, such as the hydrogen spectrum and the Compton effect. Dirac's version of quantum mechanics built on mathematical quantities that he termed 'q-numbers' and consequently his theory was often known as q-number algebra. It turned out that it was equivalent not only to the matrix version of quantum mechanics developed by Heisenberg, Max Born, and Pascual Jordan in Germany, but also to the wave mechanics that Erwin Schrödinger presented in 1926.

The exact relationship between the various versions of quantum mechanics was clarified by Dirac in the fall of 1926, in a paper 'On the Theory of Quantum Mechanics', which counts among the foundational works of modern physics. With this work, quantum mechanics was presented in a general and logical way from which the previous formulations could be considered special cases. Most physicists recognised it to be a landmark paper, but also found Dirac's condensed, algebraic theory difficult to understand. 'Dirac has a completely original and unique method of thinking', Schrödinger wrote to Bohr in 1926; and he added that 'he [Dirac] has no idea how *difficult* his papers are for the normal human being'. At about the same time, Einstein wrote to the Dutch physicist Paul Ehrenfest: 'I have trouble with Dirac. This balancing on the dizzying path between genius and madness is awful.' Indeed, Dirac's papers were usually considered lucidly written, yet very hard to understand. When Einstein visited Ehrenfest in the fall of 1926, Ehrenfest wrote to Dirac that '[we] are struggling together for hours at a time studying your work, for Einstein is eager to understand it. . . . We spent many, many hours going over a few pages of your work before we understood them! And many points are still as dark to us as the most moonless night!'

In May 1926, Dirac completed his Ph.D. dissertation and at about the same time he was assigned by Fowler to lecture on quantum mechanics to the few students of theoretical physics at Cambridge. The course, which was the first one of its kind ever taught at a British university, attracted among its students Mott, Alan Wilson, and the American J. Robert Oppenheimer; it formed the foundation of Dirac's later textbook *The Principles of Quantum Mechanics*. From the fall of 1926 to the spring of 1927 Dirac left Cambridge to study with Bohr in Copenhagen and with Heisenberg and others in Göttingen. The stay was immensely fruitful, two of the finest fruits being the transformation theory of quantum mechanics and his quantum theory of electromagnetic radiation, the first version of quantum electrodynamics. It is noteworthy that these foundational theories were developed outside Cambridge, in research environments of a kind that did not exist in Cambridge or elsewhere in England. The theories were immediately recognised to be of the greatest importance and they are still counted as among the pioneering works of quantum mechanics. For example, Heisenberg referred in late 1926 to Dirac's transformation theory as 'an extraordinary advance' and an 'extraordinarily grandiose generalisation' of quantum mechanics. Dirac's radiation theory of 1927 served as the foundation of quantum electrodynamics and initiated a new field of research that would soon occupy centre stage in theoretical physics.

In the summer of 1927 Dirac was back in Cambridge. He was by now recognised not merely as a young promising physicist, but as one of the world's finest experts in quantum theory. As an indication of his changed status in the international physics community, he was invited to participate in the prestigious Solvay conference of that year, focusing on the quantum theory of electrons and photons. At age twenty-five, he was the youngest of the carefully select participants.

A WORLD OF ANTI-PARTICLES

In spite of Dirac's early successes as a quantum theorist, at the Solvay conference his greatest contribution to theoretical physics was still to come. It was his 1928 relativistic theory of the electron, a work which

is justly celebrated as one of the great landmarks in the history of science. Dirac was from an early age fascinated by the theory of relativity, and so it was natural for him to look for a relativistic extension of quantum mechanics. The original quantum mechanics of Heisenberg, Dirac, and Schrödinger was non-relativistic, that is, valid only for particles moving relatively slowly (compared with the velocity of light).

Guided by very general principles of invariance Dirac found an equation which satisfied the basic requirements of both relativity and quantum mechanics. This so-called Dirac equation had the same formal structure as the Schrödinger equation, but mathematically and conceptually it was quite different. By following a route that was essentially mathematical, he arrived at an equation from which he was able to deduce the correct spin of the electron. Most remarkably, this important quantity was not brought into the theory, it followed from it. Dirac recalled that, 'It was a great surprise for me when I . . . discovered that the simplest possible case [of the equation] did involve the spin.'

Dirac's publication of early 1928 came as a bombshell to the quantum physicists, who immediately recognised that this was the theory they had looked for. Even before Dirac's theory was confirmed experimentally, it was received enthusiastically. According to the Belgian physicist Léon Rosenfeld, it 'was regarded as a miracle'. Rosenfeld has told how 'the general feeling was that Dirac had had more than he deserved! Doing physics in that way was not done! . . . It [the Dirac equation] was regarded really as an absolute wonder.' Dirac's theory of the relativistic and spinning electron marked the end of the first, heroic phase of quantum mechanics, and the start of a new era. It was as though the equation had a life of its own, full of surprises and subtleties undreamed of by Dirac when he developed it. During the next couple of years, these aspects were uncovered by many physicists, theorists as well as experimentalists, who found in Dirac's theory a goldmine of knowledge waiting to be digged out. Although the Dirac equation has since been modified to account for minor effects, it is still regarded as a foundation of quantum mechanics.

The greatest nuggett of gold hidden in the equation was found by

Dirac himself – Heisenberg generously called it 'perhaps the biggest jump of all big jumps in physics of our century'. In his attempt to understand the physical meaning of the 1928 equation, Dirac realised that, in a formal sense, it included solutions that referred to particles with negative energy. On the other hand, he also knew that real particles must have positive energy, and thus he was led to think of a physically valid interpretation of the solutions. What could they refer to? At the time all physicists thought that matter consisted of electrons and protons only. In 1929 Dirac came up with a remarkable solution to the enigma, namely – to put it briefly – that the proton is an electron in disguise. He assumed an infinite, unobservable 'sea' of negative–energy electrons and suggested that the protons were 'holes' in the sea, hence with positive energy. As he wrote to Bohr: 'I think one can understand in this way why all things one actually observes in nature have a positive energy. One might also hope to be able to account for the dissymmetry between electrons and protons; one could regard the protons as the real particles and electrons as the holes in the distribution of protons of [negative] energy.' In spite of Bohr's and other physicists' almost unanimous rejection of the electron–proton theory, Dirac kept to it for more than a year. He felt much attracted to it, for other reasons because it promised a reduction of the known elementary particles to just one fundamental entity, the electron. As he emphasised in an address of 1930: 'It has always been the dream of philosophers to have all matter built up from one fundamental kind of particle, so that it is not altogether satisfactory to have two in our theory, the electron and the proton.'

Dirac believed that the dream of philosophers was on its way to being realised, but after further thinking and calculations he was forced to admit that this was not the case. Yet he kept to his 'hole' picture and deftly turned the defeat into a victory by postulating in 1931 that the hole was an anti-electron, 'a new kind of particle, unknown to experimental physics, having the same mass and opposite charge to an electron'. This was a most daring hypothesis, in agreement with the principles of relativity and quantum theory, but at the time without

any experimental support whatsoever. And in his remarkable 1931 paper, 'Quantised singularities in the electromagnetic field', Dirac went considerably further. He argued that the proton would have its own anti-particle, a negatively charged anti-proton, and also that there existed magnetic charges analogous to electrical charges, so-called magnetic monopoles. A few years later, he speculated that the complete symmetry between particles and anti-particles would probably imply the existence not only of anti-atoms but of entire anti-worlds. At his 1933 Nobel lecture, he suggested the existence of anti-stars 'being built up mainly of positrons and negative electrons. In fact, there may be half the stars of each kind.'

The theory of anti-particles was at first met with considerable skepticism. After all, where were these hypothetical particles? It was only after 1932–3, when positive electrons or positrons were detected in the cosmic radiation, that the status of Dirac's theory changed. The existence of the anti-proton was revealed only in 1955, when it turned up in accelerator experiments, and in 1996 physicists could report the first detection of an anti-atom, anti-hydrogen. The fate of Dirac's prediction of the magnetic monopole has been different; the particle has not been found either in nature or in experiments. Monopole hunting became popular in the 1970s, and in a few cases premature claims were made that the particle had been found. Dirac found the theory of the monopole to be very interesting and developed it at various occasions, but eventually he came to the conclusion that monopoles probably do not exist in nature.

As an indication of Dirac's growing scientific reputation, in 1930 he was elected a Fellow of the Royal Society on the first occasion after being proposed. At age twenty-seven, he was one of the youngest Fellows in the history of the Royal Society. The year before he had been appointed a University Lecturer at Cambridge and also Praelector in Mathematical Physics, a post with nominal duties only. The culmination of Dirac's career in physics came in 1933, when he was awarded the Nobel prize – sharing it with Schrödinger – for his 'discovery of new fertile forms of the theory of atoms and for its applications'. It was no

doubt the 1928 theory of the electron and its further development to the theory of anti-particles, that contributed most heavily to his prize, but Dirac received it for his entire production since 1925. When Dirac was notified about the honour, at first he wanted to refuse it because of the publicity it would inevitably bring with it. And Dirac did not like publicity. *Sunday Dispatch*, a London newspaper, portrayed the new Nobel laureate as 'the genius who fears all women' and 'as shy as a gazelle and modest as a Victorian maid'. In spite of his mixed feelings about the Nobel prize, Dirac did go to Stockholm to receive the prize and deliver the traditional Nobel lecture.

Only two years after having completed his Ph.D. thesis, Dirac received his first offer of a chair. He was asked if he was interested in becoming Professor of Applied Mathematics at Manchester University, but declined the offer. He wanted to continue his own style of life, be free to cultivate the scientific areas he liked, and to stay out of administration and university politics. He was quite content with his position at Cambridge and over the following years he routinely declined offers of chairs of theoretical physics, both in England and abroad. Among Dirac's reasons to turn down the many rewarding offers was probably that he knew that the prestigious Lucasian Chair at Cambridge was to be vacant with Joseph Larmor's retirement in 1932. When Dirac was appointed Lucasian Professor of Mathematics in the fall of 1932, the chair was occupied by a scientist comparable to Newton, the second holder of the Chair. He remained in the chair for thirty-seven years.

PRINCIPLES OF QUANTUM MECHANICS

Although associated with Cambridge University for almost half a century, Dirac's influence on Cambridge physics was limited and mostly indirect. Dirac was anything but a school-builder of the kind exemplified by J.J. Thomson or Rutherford. His chief interest lay in fundamental physics, and he spent only a small part of his resources on teaching duties and almost none on administration. As Mott stated in 1963, 'Dirac is a man who would never, between his great discoveries,

do any sort of bread and butter problem. He would not be interested at all.' All the same, Dirac did have a very considerable impact on world physics in general and on Cambridge physics in particular, namely, through his research papers and – not least – his textbook on quantum mechanics. 'Everybody who had ever looked at books had a copy of Dirac', exaggerated the American physicist Philip Morrison. He referred to *The Principles of Quantum Mechanics*, the first edition of which appeared from Oxford University Press in 1930. Based on his lecture course at Cambridge University it was one of the first textbooks ever on quantum mechanics and, at the same time, a highly original exposition that very much expressed the author's personal taste in physics.

Principles became a tremendous success, used not only by generations of students but also studied by experienced physicists. The fifth edition of 1958 (revised in 1967) was reprinted in 1984 and is still in demand, which is highly unusual for a book of its kind. Regarded as a textbook it was and is remarkably abstract and not very helpful to the reader wanting to obtain physical insight into quantum mechanics. Based on what Dirac called 'the symbolic method', the book eschewed experiments and physical interpretations, whereas it emphasised the power of abstract mathematical methods and general concepts. It was strictly ahistorical, contained no illustrations, and only very few references. At least from a modern point of view, *Principles* was simply unpedagogical. It reflected Dirac's aristocratic sense of physics – what Born once refered to his *l'art pour l'art* attitude – and his total neglect of usual textbook pedagogy. Even experts in quantum mechanics could find it a hard read. 'A terrible book – you can't tear it apart!' Ehrenfest is said to have reacted. But other physicists found the connectedness and the coherence of the arguments to be among the book's chief qualities. In his 1931 review of *Principles*, Oppenheimer stated succintly that the book 'is clear with a clarity dangerous for a beginner, deductive, and in its foundation abstract . . . The book remains a difficult book, and one not suited only to those who come to it with some familiarity with the theory.' Yet Dirac's *Principles* is generally accepted as the most influential textbook ever in quantum mechanics.

Principles also conveyed an impression of the philosophy of science that laid behind much of Dirac's physics. For one thing, his instrumentalist view of physics – that a physical theory is essentially a mathematical structure enabling physicists to calculate experimental results – was a consistent theme of the book. In his review, John Lennard-Jones summarised Dirac's view as follows: 'A mathematical machine is set up, and without asserting or believing that it is the same as Nature's machine, we put in data at one end and take out results at the other.' The rest consisted in a comparison between results and experimental data, but without any possibility of reaching conclusions of an ontological kind. Or, as Dirac wrote in *Principles*: 'What quantum mechanics does is to try to formulate the underlying laws in such a way that one can determine from them without ambiguity what will happen under any given experimental conditions. It would be useless and meaningless to attempt to go more deeply into the relations between waves and particles than is required for this purpose.'

In addition to instrumentalism, in *Principles* Dirac also introduced another theme that would increasingly guide his research over the years to come. This was what has been called the principle of mathematical beauty, namely, the general idea that 'beautiful' mathematical equations are likely to be physically correct as well. Dirac first elaborated on the subject in a lecture he gave in 1939, and in later years he raised the idea to an almost dogmatic status.

Dirac's lecture course in quantum mechanics followed the material in *Principles* and did not change much over three decades. He delivered his lectures in a direct and lucid way, using only the phrases he found were necessary. If the students did not understand him, they could not expect much help. Dirac had only few Ph.D. students and was in general uninterested in acting as a supervisor. He seemed to lack genuine interest in his students, whom he expected to work largely on their own. When young Victor Weisskopf spent a few month in Cambridge in 1933, mainly to work with Dirac, he was disappointed. 'Dirac is a very great man, but he is absolutely unusable for any

student', he recalled forty years later; 'you can't talk to him, or, if you talk to him, he just listens and says, "yes".'

Dirac's reticence did not change much over the years and undoubtedly diminished the impact he would otherwise have had on the physics environment in Cambridge. Stephen Hawking, the famous astrophysicist who became Lucasian Professor in 1980, was a graduate student of Dennis Sciama, who himself had had Dirac as his supervisor around 1950. Although Hawking was a member of the same department as Dirac from 1962 to 1969, the newly formed Department of Applied Mathematics and Theoretical Physics (DAMTP), he never saw him. 'That was because Dirac belonged to the old school who didn't believe in these new-fangled departments of pure and applied mathematics, but worked in their college rooms. And I was working on classical general relativity and not quantum theory at that time, so I didn't go to his lectures.' In fact, when DAMTP was formed in 1959, Dirac did not accept the offer of a room in the new building. He preferred to work in his house on Cavendish Road.

Further reading

J.G. Taylor (ed.). 1987. *Tributes to Paul Dirac*. Bristol: Adam Hilger.
Behram N. Kursunoglu and Eugene P. Wigner (eds.). 1987. *Paul Adrien Maurice Dirac. Reminiscences about a Great Physicist*. Cambridge: Cambridge University Press.
Helge Kragh. 1990. *Dirac: A Scientific Biography*. Cambridge: Cambridge University Press.
R. Corby Hovis and Helge Kragh. 1993. P.A.M. Dirac and the beauty of physics. *Scientific American*, **268** (May), 104–9.
Abraham Pais *et al.* 1998. *Paul Dirac. The Man and his Work*. Cambridge: Cambridge University Press.

Paul Adrien Maurice Dirac (1902–84), born in Bristol and until 1919 a Swiss citizen. Dirac studied electrical engineering at Bristol University and physics at Cambridge University, where he received his Ph.D. in 1926 and was appointed Lucasian Professor of Mathematics in 1932. The following year he received the Nobel Prize in physics, sharing it with Erwin Schrödinger. He was elected a Fellow of the Royal Society in 1930. Since 1925 he developed the new quantum mechanics, a field in

which he belonged to the pioneers and in 1931 wrote the influential textbook *Principles of Quantum Mechanics*. His most important contribution was perhaps his 1928 theory of relativistic quantum theory, which included an explanation of the spinning electron and belongs to the cornerstones of theoretical physics. Based on this theory, he proposed in 1930–1 the existence of anti-particles and magnetic monopoles. Other of Dirac's work dealt with cosmology, in which area he developed a theory of the universe based on the hypothesis that the gravitational constant decreases over time. He also contributed importantly to the early phase of quantum electrodynamics and, from the 1950s, to the general theory of relativity. Much of his work rested on his firm belief that fundamental physics must have a 'beautiful' mathematical structure, a theme which dominated his thinking from about 1935. Dirac remained in Cambridge until 1969 and after retirement he went to Florida, where he continued working. In 1937 he married Margit Wigner, with whom he had two daughters. He died in Tallahassee, Florida, where he is buried.

Helge Kragh, Professor of History of Science at the University of Aarhus, Denmark. From 1987 to 1989 he was Associate Professor of Physics and History of Science at Cornell University, USA, and from 1995 to 1997 Professor of History of Science at the University of Oslo, Norway. His areas of work include the historical developments of physics, chemistry, cosmology and technology in the nineteenth and twentieth centuries. Among his books are *An Introduction to the Historiography of Science* (1987), *Dirac: A Scientific Biography* (1990), *Cosmology and Controversy: The Historical Development of Two Theories of the Universe* (1996), and *Quantum Generations: A History of Physics in the Twentieth Century* (1999).

17 Alan Turing
ANDREW HODGES

Alan Turing was, in a phrase, the founder of computer science. But this article will not rehearse the chronology of achievement or the claims of priority but suggest deeper questions in the motivation and culture of mathematics and its relationship to science and history. There are no definitive answers to these questions, and neither can Alan Turing be comfortably classified as a Cambridge mind; he elicited the contradictions and conflicts of that ambience.

ISOLATION AND UNIVERSALITY

Science now craves public understanding, but popularity sits ill at ease with the years of dedication to learning, challenge to received ideas, and sacrifice of advantage that science requires. Science tries to explain the universe; yet to the public (and to publishers) sits in a small specialist niche. The contradiction is even more marked in mathematics; so few, even within the sciences, can picture modern mathematical research. Recent authors have won praise for conveying the sense of struggle and devotion on the unroyal road, but have done so by omitting serious mathematical content, a musicology without knowledge of music, trying to popularise the essentially unpopular.

This creates an isolation both for mathematical culture, and for individual mathematicians. There are escape routes: one is a parochial tunnel vision, or a sort of mathematical camp, making a joke of everything. But there is a heavier burden for those who see their mathematics as the foundation stone of certainty. Turing expressed this from the beginning. An isolated schoolboy in his philistine public-school setting, Turing read Einstein and Eddington, and wrote as if they were his friends. From them, perhaps the first modern media figures in science, and perhaps also from Bertrand Russell, he acquired

the confidence to speak from the most shy and solitary position of universal themes. This confidence did not come from his family status; the Turings were anxious clingers to upper-middle-class status. It came from the Republic of Numbers.

Thus a young Turing announced the universal machine in 1936; the backroom-boffin proposed an electronic computer design in 1946; the unknown Manchester mathematician announced the prospect of artificial intelligence in 1950. But when he spoke to this effect on BBC radio in 1951, there was something the listeners could not know: unlike Eddington, Einstein, and Russell this unknown theorist had been stupendously useful.

THE USES OF NUMERACY

Alan Turing was a Cambridge student from 1931 to 1934, a Fellow of King's College between 1935 and 1952. If anywhere was Turing's territory, it was the Keynesian ambience at King's; and indeed as a Cambridge mind Turing had a little in common with Keynes: living a liberal, private life, drawn to pivot of worldly affairs; yet all the time, as a homosexual, one of its outcasts. But Turing was not a product of Keynesian high culture nor a conspicuous success within it. Slightly closer, perhaps, was the Cambridge mind of G.H. Hardy: the very private yet the most public in saying what many would not have dared say in 1940 with his essay, *A Mathematician's Apology*.

Hardy's sharpest reproof was to utilitarian – in his time marxist influenced – talk of mathematics for planning and order. (It reads oddly now. Today's emphasis would be on short-term profit, which never entered into Hardy's dialectic.) It is now a commonplace remark that Hardy was wrong about relativity and quantum mechanics having no military use, as was soon to be shown. Hardy would have been surprised, I think, that advanced number theory is now of great significance for commercial cryptography. Though allowing the possibility of such an application, Hardy still ended ringingly 'I have done nothing useful.' Turing made no such predictions of purity and issued no soul-searching apology.

FIGURE 17.1 Alan Turing.
Source: Turing Archive, King's College, Cambridge

But Hardy's point was essentially a *moral* one about utility being irrelevant to value: the aesthetics of Wilde. He denied the justification by utility because it was untrue: not the real motivation for real mathematical thought. In this awkward truth telling, Turing also shared Hardy's integrity of integers. 'Phoney' was one of his favourite words of opprobrium, along with 'Politicians, Charlatans, Salesmen'. Turing's motivations, even his practical work, were as unrelated to public or private profit as Hardy's.

It is curious to think whether Turing's obscure and unfashionable

work of 1936 would receive a grant by modern criteria, which encourage scientists to make their work show immediate commercial relevance. Turing did in fact describe his ideas to his economist friend David Champernowne, who was appropriately sceptical of their practical value.

But Turing had also another mathematical magic, one that Hardy only touched upon as a secondary characteristic. Within the gamut of mathematics lies its strange special harmony with the physical world; hence follows a prophetic role of mathematics as the vanguard of science. Like Newton, Gauss, Riemann, Hamilton, and Hilbert – doing uneconomic work that took centuries to develop, differential geometry before relativity, complex vector spaces before quantum mechanics – the universal Turing machine came before electronic computers.

Turing had this prophetic gift, and his work moved from the pure and abstruse to world-conquering utility. Turing's 1936 work took the logical puzzles of self-reference, developed by Gödel from the work of Cantor and Russell, and by expressing it in the language of calculation, found both the absolute limits of computability, and the concept of the universal machine. His paper spoke to a small subset of pure mathematicians. Yet within ten years Turing turned the universal machine into the practical invention of the computer in its modern sense. Turing codified operations on numbers *by* numbers, and thereby both solved a deep problem in mathematics, and identified an idea essential to the modern world. Computer programs, operating on data, are themselves data; and Turing was eager to put this idea into application as soon as it could be embodied in electronics.

That application required the war. And war means loss, for mathematics as for everything else. Hardy's 'useless' relativity and quantum mechanics left a Cold War legacy of escalating arsenals; so did Turing's logic.

MATTER AND SPIRIT

Cambridge mathematics has been marked in its separation of 'pure' and 'applied' cultures. When in 1954 Max Newman began his

'Biographical Memoir' of Turing, the question to which he first turned was that of which side of the divide Turing belonged, pronouncing him as being at heart an applied mathematician – a judgment that must have surprised many who saw Turing as a pure logician. Turing himself never referred to this distinction; he called himself 'a mathematician' and applied himself anywhere within the logical and the physical worlds, especially at their interface.

I have described him elsewhere as natural philosopher rather than mathematician, for Turing's inspirations had a first base in chemistry, the physics of mind and matter, in which the mind of Eddington and the matter of his first love, Christopher Morcom, combined. Cambridge mathematics was the right vehicle for this train of thought; and Newman, topologist and true Cambridge mind, was the catalyst who brought Turing to a frontier in the *Entscheidungsproblem*. But it was not Cambridge that taught Turing to treat logic as applied mathematics; that came from something deeper. As Newman wrote in 1955, his introducing paper tape into logic came as a shock; perhaps partly because it was a mundane technological image sullying high mathematics. But, unlike Hardy, Turing did not distinguish classes of mathematical work. In 1937, while absorbed in his most abstruse work, the development of 'ordinal logics' to probe and classify the uncomputable, Turing was also thinking about applying logic to ciphers, and building a cipher machine with electromagnetic relays.

Turing applied unsuccessfully for a Cambridge University lectureship in 1938; the University never employed him.

DIE ZWEITE HEIMAT

Turing's local home was at King's, his global home was that of Hilbert and Hardy for whom the world was a single country. Bletchley Park was a home from home, linked strongly to King's College culture, but housing a world intelligentsia. But Turing, unlike Hardy, was prepared to do the 'trivial' mathematics of the Enigma. As if to cheek Hardy, he made chess playing his analogy for intelligence, when starting discussions of artificial intelligence in wartime huts.

But there was 'real' mathematics too: in group theory and probability, Turing brought the power of mathematics to a pre-scientific world (where only the Polish algebraists had gone before) and created his own information theory. To break the Enigma, Turing and Welchman put their astonishing logic into the electromagnetic Bombes. At the same time Turing took on Naval Enigma alone, disregarding the discouragement of his superior, Commander Denniston: 'You know, the Germans don't mean you to read their stuff, and I don't expect you ever will.' And why did he? His colleague Hugh Alexander's report, released from secrecy fifty-five years later, reveals that Turing said they *had* to be broken because the intellectual challenge was so great. Broken they duly were. Hardy-like, he refused to say he had done it from the call of duty; the truth was that he had found it fascinating. I have always found the greatest drama in this, that the innocence of deeply unworldly mathematics met the call of the greatest world crisis, and met it at its very centre.

Turing was fully aware of international events; he had even sponsored a refugee. But his Cambridge scientific mind made the war a chess game, one we now know to have been a duel with the young German logician Gisbert Hasenjaeger, entrusted with the keys of the Reich. After losing that war Hasenjaeger returned to logic, and when eventually interviewed on television, had outlived his victor by forty-five years. In 1936–7 Turing portrayed cryptography as something that would flow from his logic, something that would be a game against Germany, and something that meant an essentially moral choice, a sacrifice of purity. He chose in 1938, as Snow White bit on the apple. Did he sense even then, as he signed the Official Secrets Act, that he was killing truthfulness? It is strange irony that Turing's magical design for the Bombe turned upon the concept of following through the proliferating implications of false hypotheses.

The Bombe required a novel synthesis of ideas and engineering; and an ingenuity of logic that few at Bletchley Park understood; it was a miracle that Denniston was persuaded to invest so much in a great gamble. The first Bombe to be delivered was named *Agnus* by Turing: a

joke that atheist Hardy might have made (though Hardy at that moment was making his unapologetic apology, trying to avoid thought of the war). Further irony showed in the names they used: *Turingismus* as if this ingenious statistical analysis of the Lorenz machine ciphers were some Hegelian philosophy, *ROMSing,* as if the placing of long paper tapes against each other were the *Resources of Modern Science* of marxist planners. *'Cillies'* were silly German Enigma operators' errors; but also perhaps the silliness of spending scientific talent on the crimes and folly of mankind. (Now, exiles from mathematical physics who enter the ephemera of futures markets, computer operating systems, and e-commerce could echo the sentiment.) Newman, at Bletchley after 1942, lamented that it could not have been a pure mathematical research group.

But it was science; unlike that shown in television's *Station X*, it was the power of scientific method; the production line of information, with Turing uniquely placed to see electronics making program handling practical, and so able to embody the universal machine. Turing saw the future far clearer than practical people, having borrowed from it to defeat Nazi Germany.

PRIVATE PLANS AND PUBLICATION

In 1945 British vision for planning was well abreast of the United States, with a confidence in turning war work to future prosperity. Turing evoked the potential of the computer in his plan for the Automatic Computing Engine at the National Physical Laboratory (NPL), his universal machine turned to national utility. So did Newman, taking Turing's idea to Manchester for a computer devoted to pure mathematics. But Turing and Newman both lost their pitch. The turning point was in 1948, as Turing gave up on the NPL, and joined Newman and Blackett as Cambridge minds at Manchester. Newman had secured a Royal Society grant for the development of a computer – regarded as a universal Turing machine, as the terms of Turing's appointment in 1948 made clear. But Turing was unable to enjoy more than superficial collaboration with the engineers who had

taken over the project. The Royal Society's priorities were forgotten: the Manchester computer was engineered for the British atomic bomb, everything Hardy might have feared in his grimmest lines. The successful synthesis of ideas and engineering in that other, defensive, logical Bombe remained Turing's secret. And later American commercial success soon ensured that the history of the computer was located in the United States.

Turing never clearly specified or published his claim to the stored program concept. Untroubled by Research Assessment citation counts, he left that problem to historians of the computer. Newman probably never thought of the computer as real mathematics, and might well not have thought its central principle worth publishing. It is more of a puzzle that Turing, quite prepared to pursue trivial mathematics, did not elucidate the way that the principle of 1936 had turned to the practice of 1946.

But Turing's computer was neither the useless reality of Hardy nor the useful reality of the atomic bomb; it was to do with simulating the brain, the arena for experiment on his thesis that the operation of the mind is computable. Even in his practical report for the NPL in 1946, this priority had shone through. As I have argued elsewhere, it was probably in 1941, in what must have seemed a revelation, perhaps enhanced by seeing machines overtake aspects of human guessing, that he decided intuition and originality must also be computable processes, explicable by programs allowed to change through experience into forms not originally written down. By 1945, Turing had concluded that the uncomputable was not needed to account for intuition or creativity, and thereafter argued vigorously to this effect. It is remarkable that Turing who showed such originality, denied originality its apparent meaning; he claimed intuition was all learning by experience. He was an anti-Socrates, though his life led to a Socratic end.

Turing set forth a major scientific paradigm as a result, by opening the arena of the discrete state machine. This at least he did not throw away; and it has given him his lasting status as a philosopher as well as mathematician. He transformed the ancient mind–body ques-

tion by connecting it with computability, and turned it into a new experimental science. He was not saying that the brain is like a computer in its architecture; rather, as he often explained, that if the operation of the brain was computable, then, no matter how complex, it could be simulated by a program on a computer. For this was the well-proved property of the universal machine. And this was the theme that drew out his famous 1950 publication.

It was overconfident in its fifty-year prediction, as in 2000 we know; but we see now an overconfidence in all the cybernetic pioneers, the optimism perhaps essential to all inspired research, depression being its inevitable corollary. But this still leaves a puzzling aspect to Alan Turing: he failed to follow up or publish his neural net sketches, the learning machines whose potential was fundamental to his vision of intelligent machinery. There was a self-defeating element; a dark side to the vision. His ideas remained unknown when the American school of thought developed after 1956. Likewise he was lost to the history of practical computer development, never publishing his programming ideas.

LYING WITH MEN

A possible answer to the puzzle is that whatever the power of inspiration, and however strong the individual, the infrastructure of human support and communication is necessary for effective creation. At Bletchley Park, although preferring to have problems to himself, Turing was nonetheless in a group. He had to accept the value of Welchman's input to the Bombe, something he had missed; had to give and take with Alexander, accept American collaboration, and much else. After 1945 his isolation was reasserted; support never came from Cambridge (King's College could not support any computer-building ambitions), and only partially, through Newman, from Manchester. Turing was prone to depression; running to relieve the stress not of mathematics but of implementing it in the so-called real world. (Turing developed this into running marathons to Olympic standard.)

Another possible answer lies in the Wilde side: the demoralising

burden of trying to live an honest criminal life. *For each man kills the thing he loved,* he told Joan Clarke in 1941 when he rejected the temptation to live a lie, rather more dramatically than indicated in the staged and televised *Breaking the Code.* His gay identity gained more support and confidence after the war, and what he had seen as a curse in 1930s, he began to see as a frontier of exploratory consciousness. It was the wrong time in the wrong place for that. Once arrested, Turing was turned into class traitor and Cold-War risk in the moral panics and national security crisis of 1952. Amidst this, for him the issues of truth and trust were paramount. *Turing believes machines think, Turing lies with men, Therefore machines do not think,* he wrote, camp humour combined with mathematical camp, but with truth at the centre. It was his honest way of dealing with the intensely dishonest world of the worldly, though with another irony, Turing was found out through his telling a lie to the police. Cambridge was little help to him in this; he was too open and democratic to be a Cambridge mind. But Newman was his character witness and called Turing 'particularly honest and truthful'. After the initial lie, Turing acquitted himself with Hardy's honesty: because, as Hardy said of the primality of 317, it is so.

Unlike Hardy, Alan Turing was on the receiving end of war through his usefulness; it made his brain a repository of the most secret knowledge of the West, and a security risk when he took that brain to Norway or Greece. Such provocative adventure was not the same as Hardy's escape route, the cricket field. But both must have known the same soul-destroying burden of being counted as less than human: as unbearable perhaps as trying to implement mathematics that no-one understood, or knowing that salvation had come from that which society hates and fears and concerning which one must remain completely silent.

Newman was described as 'a professor of pure mathematics' in the local newspaper's criminal reports. And so he was: just another professor. Newman and Turing, like Hardy, had an entirely realistic view of their worldly status. Richard Feynman, who nearly overlapped with Turing at those stifling Princeton tea-parties, had the fortune to live

into times where such conflicts could be expressed more easily. *Why should he care what other people think* could be asked as well of Turing as of Feynman, and again like Hardy, who in C.P. Snow's summary 'just didn't give a damn'.

Turing, like Feynman, could be accused of a higher selfishness along with high mathematics. In his last days Turing insisted on a social ideal of free association quite incompatible with his secret state status, rather as Feynman felt himself 'the only free man' who could speak up for his Pasadena low life. But Turing was not a free man. His last postcards, in March 1954 referred mysteriously to electrons not allowed to associate too freely; another maths–camp joke, given meaning by the dark rider, that it was laid down for the good of the electrons themselves. The Agnus Dei, in Christian myth, is sacrificed by his own side.

SLEEPING DEATH

In the wake of Turing's dramatic suicide in 1954, the task of interpretation fell first to Newman, writing the 'Royal Society Biographical Memoir'. This document is about Turing; but also about Newman and his Cambridge mindedness. Newman had no interest in convention or pleasing worldly people. Yet integrity and truth are highly divisible in social practice, and the 1952 trial was no more mentionable in British obituaries than it would have been in Russia. It is perhaps impossible now to recover how shocking Turing was, blasting through public and private truth in a world where no-one was ever supposed to question bishops, judges, or police. 'The poor sweeties', Turing called the police watching him in 1953. I found in 1977 that this traumatic story was undiscussed memory in all who knew him, like some frozen mammoth from the Cold War crisis years. After completing the ghastly task of this 'Memoir', Newman rarely spoke of Turing, and the world largely followed his example.

In that 'Memoir', Newman tried to express an overall unity in Turing's thought, which subverted the conventional separation of 'pure' and 'applied'. But Newman was apparently embarrassed by

Turing's work in the trivial mathematics of computing. He misrepresented Turing, referring to designers of computers not knowing of Turing's universal machine concept, apparently forgetting that Turing was one such designer himself. (This omission did lasting damage to Turing's reputation.) The now-popular Turing Test was perhaps what Newman meant by disparaging the 'gimcrack' quality in Turing's thought. (I wonder what Newman would have thought of today's exaggerations and promotion stunts in name of 'public understanding of science'; its science exhibitionism at the expense of science research.) Newman also gave a misleading picture of Turing's work at Bletchley. He was of course bound by intense secrecy, but, even so it is strange that he rendered the invention of the Bombe, the fight with the Service mentality, the night shifts breaking daily Enigma settings and hair-raising voyages, as a happy time of 'congenial colleagues' and 'a mild routine to shape the day'.

Newman lamented that the trivial demands of war took Turing away from serious problems, and gave far more attention to the ordinal logics of 1939 than to mere computers. Turing's real mathematics, he suggested, had been eclipsed by mundanely useful chess playing. In so saying, Newman's survey evoked Turing's fertile thought-world of 1938, the would-have-been problems he could have taken on instead of breaking the Enigma and inventing the computer: perhaps pursuing the Riemann zeta-function, perhaps the extension of ordinal logics. The justification for Newman's unpopularisation is that to understand mathematicians you must understand this priority and scale in time. Their unworldliness may turn out in time to be the most realistic.

Roger Penrose has recently expressed this theme: though holding Turing the leading pioneer of the dominant technology of the late twentieth century, he has stressed that Turing's own work showed the limitations of computability. He has also pointed to a real question here. What would Turing have made of the physics of the brain in 1938, if he had taken seriously the idea expressed in his ordinal logics that the mind does something uncomputable in seeing 'intuitively' the truth of

a Gödel statement? We are entitled to ask because of Turing's earlier fascination, stimulated by Eddington, with the quantum mechanical physics of the brain and the 'nature of spirit'. There is now more than one fictional Turing set in 1939, but a more interesting story lies in this unpursued investigation. What if Wittgenstein in 1939 – but this remains a contrafactual enigma.

THE CHILD IN TIME

When Turing was elected a Fellow of the Royal Society in 1951, the citation referring to the work he had done fifteen years earlier, he commented that they could hardly have elected him at twenty-four. It was Hardy's comment that maths is a young man's game, though the point of this observation is often overdone, given that Hardy said he was at his best at just after forty. Newman wrote of Turing as being at the height of his powers in 1954, when he was forty-one.

Turing proudly acted the *enfant terrible.* Nowadays, postmodern critics pick over the Turing Test for its alleged agenda of gender identity, but if Alan Turing had an identity crisis it was not over gender but over innocence and experience: the roles of boy and man. He demanded freshness, and one reason why he never gave definitive completions of his work was that he always wanted to start on something new. Excluded from the real men's computer engineering world, his late morphogenetic theory, which seems to have gone back to childhood wonder, gave him a way to make fresh discovery on the computer he had invented. The origin of that work lay in D'Arcy Thompson, back at the beginning of the century; and in other ways too, Turing's work sprung from the crises of 1900. The germs of Turing's work were to be seen in Hilbert's 1900 problems, Russell's 1901 paradox, and Planck's 1900 quantum. What Turing led was not so much a linear advance in any one of these fields as allowing their synthesis, and connection with the practice of computation and communication. Turing's work has spanned much of the century, and in 2000 quantum cryptography is the first practical advance outside the scope of Turing's conceptions.

The stone that the builders reject may turn out to be the cornerstone. Russell thought his paradox a blow, not a discovery. But mathematics works a slow miracle, not the science magazine hype which demands a revolution every week: it took centuries of logic to see what Gödel saw, and, as Turing wrote, one man can do so little in a lifetime. Turing never foresaw quantum computing; did not see the advent of complexity theory. All this was left to others, Feynman amongst them, but also Roger Penrose, who has sought to repair Turing's lack of investigation into the physics of computability, which goes to the heart of Turing's connection between the logical and physical. At the very least, Penrose has clarified the assumption, implicit in Turing's 1950 argument, that the physics of the brain is computable.

Turing made a young man's game of his assumption of the computability of the mind, by framing his famous Test rather as he had made a game of the Second World War. It is nonetheless, a deadly serious question, as central to science as that war is to history. It is the question of Frankenstein, the creation of real life; a question so great as to be unbearable for an individual to carry, unbearable perhaps as being critical to the war. It will take the new century, perhaps more, to see whether Turing is right, or whether Penrose has located that awkward detail, rejected by the builders of Artificial Intelligence, which becomes cornerstone of an entire new world of thought.

Both Turing and Penrose learnt about quantum-mechanical reality from the lectures of Dirac, another mind hard to classify as Cambridge's. There are mysteries to Turing's last year; we do not know what the secret state demanded; we can only guess whether Turing may have been trying to regain the lost freshness of Cambridge youth in 1935. There is a clue to Turing's motivation: in 1951 Turing gave a radio talk which referred to Eddington and in two sentences said that quantum mechanics might mean that the behaviour of the brain was not computable. So Turing was alive to this question – Penrose's question – and it may well be that this is why he thereafter concerned himself with wave-function reduction, though not coming to any conclusion before his death.

Hyperboloids of wondrous light, wrote Alan Turing to his friend and student Robin Gandy in a strange last postcard of March 1954 'from the unseen world' – an allusion to Eddington, but uncanny in foreshadowing the wondrous light of Penrose's relativity, uneconomic as Hardy, unworldly as Newman, unpopular as Turing himself.

Alan Turing killed himself on 7 June 1954, but for many people his death has brought hateful mathematics to life. What was for decades an unspeakable truth, has become the small change of Internet newsgroup banter. Now, talk of memorials to him reinforce the fact that the most awkward and unpopular can become the cornerstone; I myself would say fit for a cornerstone in Trafalgar Square, a non-violent Nelson on patient service for a new century. He took away the sins of the 1940s world, then found himself the sinner. But was it really his sexuality that demanded the public stoning? Or the transgression of demarcation lines, lines of class and trade? Or did it lie in being so necessary in 1940 when the British state was humiliatingly dependent on individual wits, wits that the literature-educated could not fathom? Or did his sin lie in the unbearable duty of genius to speak the truth? His theory denied the Self; yet he made a self-sacrifice of the wild scientific mind.

Further reading

The Collected Works of A. M. Turing (ed.) J. Britton, P.T. Saunders. D.C. Ince, R.O. Gandy and C.E.M. Yates, 4 vols. (1992–2001). Oxford: Elsevier.

A.M. Turing. 1950. Computing machinery and intelligence, *Mind*, **59,** 433–460. Reprinted many times, e.g. in Margaret Boden (ed.), *The Philosophy of Artificial Intelligence.* Oxford: Oxford University Press.

F. L. Bauer. 2000. *Decrypted Secrets,* Second edition. New York: Springer-Verlag.

Arthur S. Eddington. 1928. *The Nature of the Physical World.* Cambridge: Cambridge University Press.

G.H. Hardy. 1940. *A Mathematician's Apology.* Cambridge: Cambridge University Press.

F.H. Hinsley and Alan Stripp (eds). 1993. *Codebreakers.* Oxford: Oxford University Press.

Andrew Hodges. 1983. *Alan Turing: The Enigma.* London: Burnett; New York: Simon & Schuster. New editions London: Vintage, 1992; New York: Walker, 2000.

Andrew Hodges. 1997. *Turing: A Natural Philosopher*. London: Phoenix. Also New York: Routledge, 1999. Also in *The Great Philosophers*, eds. Ray Monk and Frederic Raphael. London: Weidenfeld & Nicolson, 2000.

Andrew Hodges: Alan Turing Website at http://www.turing.org.uk

M.H.A. Newman. 1955. Alan Mathison Turing. *Biographical Memoirs of Fellows of the Royal Society*, **1**, 253–63.

Roger Penrose. 1989. *The Emperor's New Mind*. Oxford: Oxford University Press.

Roger Penrose. 1994. *Shadows of the Mind*. Oxford: Oxford University Press.

Alan Turing was born in London on 23 June 1912, and educated at Sherborne School. After a slow start he became a mathematics student at King's College, Cambridge, and was elected Fellow of that college in 1935. In 1936 he completed his paper *On Computable Numbers, with an Application to the Entscheidungsproblem*, which simultaneously settled a problem in the foundation of mathematics, and founded modern computer science with its definition of computability and the universal machine. Turing pursued several strands of mathematical research thereafter, with two years at Princeton, but, from late 1938, work on the German Enigma problem predominated. He became the first and the chief scientific figure in the successful British cryptanalysis of German military ciphers based at Bletchley Park. His wartime exposure to electronic technology encouraged him to design for the National Physical Laboratory a first computer in the modern sense, as a practical version of his theoretical universal machine. Disappointed with the prospects there he moved to the Manchester computer project in 1948, and, while working on programming, set out the conceptual base of artificial intelligence in a classic paper of 1950. He then formulated a new mathematical theory of morphogenesis, and pioneered the use of the computer in its investigation. This, and other new interests in fundamental physics, was interrupted by his arrest in 1952 for an affair with a young man. The humiliation of enforced medical 'treatment' was compounded by loss of security clearance. His resilience and humour in the face of adversity was such, however, that his suicide by cyanide poisoning on 7 June 1954 came to general surprise.

Andrew Hodges published the biography, *Alan Turing: The Enigma*, in 1983 and his shorter *Turing: A Natural Philosopher*, appeared in the Great Philosophers series in 1997. His main research work since 1972 has been in mathematical physics, applying Roger Penrose's theory of twistor geometry to the description of fundamental particles and their interactions. He is a Lecturer in mathematics at Wadham College, Oxford, and is on the World Wide Web at http://www.synth.co.uk/

18 Francis Crick and James Watson
ROBERT OLBY

The twentieth century has witnessed a remarkable transformation in the scientific status, economic importance, and public visibility of biology. The new knowledge that lay behind this transformation was based upon scientific achievements that were made in research institutes and universities in many parts of the world beginning in the fifties. A major foundation for such success was laid when Francis Crick and James Watson, working at the time in Cambridge, suggested that deoxyribonucleic acid or DNA – the stuff of our genes – is constructed of two helical long-chain molecules wound around each other – the now-famous Double Helix. Their brief paper describing the structure in the journal, *Nature*, in 1953, has since become one of the most famous in the history of twentieth-century science.

The Double Helix, its structure suggestive of its function as repository of the specificities or 'information' of our genes and of the molecular mechanism by which the genes are duplicated, supplied the foundation upon which could be built our understanding of genetics at the molecular level. It clinched the debate already under way in favour of the nucleic acids rather than the proteins as the substance of the genes, and inspired the search for the hereditary codescript in which the genes are written. Along with the revelations of the molecular structure of the proteins, this new knowledge has made possible the development of molecular tools with which to manipulate the genetic material for the benefit of medicine and agriculture, diagnostic tests to guide preventive medical advice, and an identity test – DNA fingerprinting – to aid forensic science.

In the half-century following their discovery, both authors have continued to contribute to the science they love. Watson has gone on to build a highly visible career as an educator through his textbook *The*

Molecular Biology of the Gene, as an autobiographer with his immensely successful book *The Double Helix* (to which he has just added *The Passion for DNA*), and as an institution builder who rescued the venerable Cold Spring Harbor Laboratory on Long Island from decline. Recently he has been a force behind America's Human Genome Project as the first director of the Center for Human Genome Research of the National Institutes of Health.

By contrast Crick has continued to devote himself to research. At Cambridge he worked with the South African scientist, Sydney Brenner, to contribute to the elucidation of the genetic code, and in America with Alexander Rich to establish the structure of collagen, and with Leslie Orgel to enter the debates on the origin of life. But his most influential contribution has been his lecture to biologists in 1957 on the subject of protein synthesis. Here he laid out a set of principles which proved fundamental to the emerging science known as molecular biology. Moving to the Salk Institute in 1977 he turned to neuroscience. By his enthusiasm and magnetism he has helped to build a neuroscience community of researchers in La Jolla and to establish a rewarding collaboration with the California neuroscientist, Christof Koch. By his stimulating and speculative papers on the nature of consciousness and the function of dreams, and his highly readable overview of researches relating to attention and consciousness – *The Astonishing Hypothesis* – he has acted as a catalyst bringing researchers from diverse fields together and drawing the attention of a widening public to the neurosciences. He, too, has written his autobiography – *What Mad Pursuit* – in which he reflects on the lessons that life teaches us about how to do good science.

It was in 1951 that Francis Crick and James Watson found themselves in the same laboratory in Cambridge. Britain was still in the immediate post-war phase of recovery. Its hopes for the technical and commercial future of the country were set on nuclear power and commercial jet propulsion. While the world's first computer was left to gather dust as a military secret, the lion's share of research funds went to the applications of nuclear fission – both explosive and controlled.

FIGURE 18.1 Francis Crick and James Watson.
Source: A. Barrington Brown Science Photo Library.

Fortunately, concern expressed within the Royal Society during the war that attention be paid to the funding of fundamental research in peacetime, led to initiatives being taken in 1945 to attract direct government funding. A number of Fellows of the Royal Society who had won their laurels in the applications of science to military needs, reckoned that to move British science ahead it would be necessary to appeal to the government for funds to establish research institutes or units within or in proximity to universities, but not controlled by them. A number of research institutes funded by the government had already come into being as a result of the Development Act of 1909 and

of the Research Association formed following the First World War. Several were clustered in and around the city of Cambridge. These provided the university with scientific expertise and on-going research, unhampered by the needs of a teaching role.

In post-war Britain this strategy was to prove especially important for hybrid subjects, like oceanography and biophysics, which did not fit neatly into any one academic discipline. In the case of biophysics it was the Medical Research Council in 1947 that stepped in to fund the first unit of this kind, not in Cambridge but in London. This was Sir John Randall's Biophysics Unit, attached administratively to the University of London through the Physics Department of King's College. Randall's success was followed by that of Sir Lawrence Bragg who had applied to the same Council for funds to establish a 'Unit for the Study of the Molecular Structure of Biological Systems'. Both units were supervised by the Council's Biophysics Committee, the term traditionally used to signify the application of physical techniques in biology and medicine. However, the concentration of the Cambridge Unit's researches upon the structure and functional relations of proteins and in due course upon the sequencing of the proteins led, as we shall see, to a change of title in 1958.

It is thus no accident that the studies most relevant to Crick and Watson's elucidation of the double helix were made at these two MRC Units. The London unit began with an eclectic set of approaches but all focused upon the dividing cell. Included was the study of molecular structure by X-ray diffraction techniques. Applied by Maurice Wilkins to DNA, this programme was subsequently taken up by Rosalind Franklin, who provided most of the crystallographic data on DNA, known to Crick and Watson in its general features, that proved to be crucial to the solution of its structure.

Meanwhile, in Cambridge the research programme was focused on the three-dimensional structure of the proteins. The members of the unit were Max Perutz, joined by two doctoral students, first by Hugh Huxley and then by Francis Crick in 1949. Also working on protein structure was John Kendrew, but supervised by Dr Taylor in the

physics department. The founder figure of this tradition was the brilliant John Desmond Bernal. He had moved from Cambridge to London in 1937, but left behind the Viennese chemist, Max Perutz, who was using X-ray diffraction analysis to seek the structure of haemoglobin. Supported initially by the Rockefeller Foundation, Perutz's work became the nucleus of the programme in the post-war period around which the Unit was established.

In the 1930s the solution to the structures of molecules as large and as complex as haemoglobin was considered impossible by existing means, but the hope had emerged that by the use of what is known as Patterson analysis, certain problems could be overcome and an inductive approach to the answer obtained. Assuming that the long chains of this huge molecule were packed together in a regular manner like logs in parallel rows and layers, Perutz was hopeful. Then Crick as a Ph.D. student, gave a talk entitled 'What Mad Pursuit', in which – as he recalls it – he laid bare the inadequacies of most of the methods being used and the implausibility of the molecular structures envisaged. This behaviour was not expected of a doctoral student, but, like Kendrew, Crick was not just a doctoral student. At thirty-three years of age and a former naval officer, he was assured and used to making judgments and speaking his mind. Moreover, prior to his military work, he had nearly completed research for a doctorate in physics, which had already won him a prize at University College London, before a bomb destroyed the apparatus.

Although the war had interrupted most academic research in the sciences, it exposed young researchers like Crick to the discipline and the practical goals of military research. Moreover, in breaking the continuity of his career, it offered him the opportunity to rethink his future. The life of a research scientist in the Admiralty was attractive and well paid compared with academia. Consequently Crick was not convinced that he wanted a university career. He found he could stay on at the Admiralty in a permanent position, but, deciding against this, he explored possibilities in industrial research. Finally, it was fundamental research that won his vote. Top of his list of possible topics were

the mysteries of the origin of life and the nature of consciousness. He chose the former because it seemed closer to the science he knew. It was almost three decades later that he turned to the latter topic.

Crick's entry into biological research began in 1947 when he entered the Strangeways Laboratory, home to many and varied studies of the cell. His experimental studies apart, this was the time of his voracious immersion in the literature of biology. Two years later he was accepted into the MRC Unit at the Cavendish laboratory to undertake doctoral research on the structure of the proteins. Now he immersed himself in the methods for determining molecular structure from the X-ray diffraction patterns produced by protein crystals. Two years passed by during which Crick experienced many problems with his experimental work, but he did make two important contributions, one of them with William Cochran and V. Vand, giving the form of the X-ray diffraction pattern produced by a helical molecule in 1952. The other was his use of this derivation – known as the Fourier transform of a helix – to calculate the diffraction pattern that a helix would produce if a second coil of a higher order were imposed on it, so that it became what he called a 'coiled coil'. The result accounted for the puzzle of the most prominent feature of the diffraction pattern of the much studied protein, α keratin.

In the autumn of 1951 James Watson, then a post-doctoral student aged twenty-three, came to Cambridge in his search for a home in which to learn what he needed to know to seek the structure of the genetic material. His studies at Chicago and his research at the University of Indiana had equipped him as a microbial geneticist, but had not prepared him for the structural chemistry of the proteins. However, his inability to aid the Cambridge scientists in their structural studies meant that he could devote most of his time to studying the literature, and to discussions with Crick, the one person in the Unit who, like him, believed that, although enzymes are important, genes are even more important since they *make* enzymes. Now enzymes are proteins, and their remarkable specificity of function and diversity of structure justified their reputation among biologists and medical

scientists as the most important class of compounds associated with life. The focus of the Cambridge group on them was thus entirely justified. Naturally it was held that the gene – the determinant of a trait – is a special kind of enzyme that makes more of itself as well as making a gene product. From the 1940s onwards, however, evidence began to emerge that the genes are not composed of protein, not even of a compound of protein and nucleic acid, but of nucleic acid alone. If so, the key to the mechanism for both the duplication of genes and the synthesis of gene products – the proteins – should be sought in the chemistry of the nucleic acids. On this point Crick and Watson thought alike. Thus arose the collaboration between them and the story, so well and frankly told by Watson, of their initial failure in 1951 and their ultimate success in 1953.

Twelve years younger than Crick, Watson had completed his doctoral thesis at the age of twenty-two. But he was diffident in manner, his social graces were few, his words brief, and his curiosity highly focused upon genetics. His father had won him for 'birding', but by the time he was ready to become a post-graduate, ornithology lost out to genetics. Watson has stressed that he was no child genius, but from his father he had learned early to love reading. The University of Chicago with its Great Books courses taught him to turn to primary sources, and to concentrate on putting facts 'together in some rational scheme'. The result was that before he met Crick he had been acquiring the mental habits that would make him acceptable to his new-found friend. Important, too, was Watson's doctoral research which gave him hands-on experience with the genetics of bacterial viruses (bacteriophages or 'phages'), for, in probing the finer structure of the genetic material, geneticists were turning from the fruit fly (*Drosophila*) to the minute phages that parasitise the colon bacillus, *Escherichia coli*.

Not that this collaboration thrived simply due to the complementary expertise of Watson the biologist and Crick the physicist. True, Crick was steeped in the X-ray crystallography of the proteins and Watson in the experimental researches on phage. Watson learnt his crystallography from Crick, and Crick was introduced to the circle of

phage researchers through Watson. Through Crick, Watson was able to get to know Crick's long-time friend, Maurice Wilkins working at the MRC Biophysics Unit in London. But Watson and Crick 'hit it off', recalled Crick, 'because of a certain youthful arrogance, a ruthlessness, and an impatience with sloppy thinking', attitudes that 'came naturally' to both of them. They also developed 'unstated but fruitful methods of collaboration . . . If either of us suggested a new idea, the other, while taking it seriously, would attempt to demolish it in a candid but nonhostile manner.' This, recalled Crick, 'turned out to be quite crucial'. Consequently there was much discussion and both serious and playful argument – so much so that it was with pleasure and relief that when a second room became available for the Unit, the others happily assigned it to Crick and Watson.

In their upbringing, despite national differences and the relative poverty of Watson's family, there were similarities. Both boys sought activities to escape from religious services – Crick playing tennis, and Watson birding. Both were considered obsessive in the manner they pursued their interests, Crick in science and Watson in birding. While Crick had the benefit of an excellent secondary education at the prestigious Mill Hill School, Watson's horizons were broadened by 'the intellectual obstacle race' as he called the famed Great Books course at Chicago. Although it was the Chicago geneticist, Sewell Wright, who decided Watson's future object of study should be the gene, he had first been attracted to this subject by his reading of Erwin Schrödinger's little book: *What is Life?* (1944). It taught him that 'the gene, being the essence of life, was clearly a more important objective than how birds migrate' – the goal of his scientific enthusiasm hitherto. Crick also read this book and was impressed, for it 'expressed in an exciting way the idea that, in biology, molecular explanations would not only be extremely important but also that they were just around the corner'.

Sure enough, before Watson and Crick met, a harbinger of that promise appeared in the form of Linus Pauling's proposed structures for proteins – in particular the helical model to which he gave the name 'alpha helix'. Now proteins are huge molecules composed of long

chains of amino acids held together by what are called 'peptide bonds'. Hence the chains are called 'polypeptides'. Just how these long chains are constituted and packed into a complex protein like haemoglobin had long been debated, but it had been customary not to question certain assumptions following from the rules of crystallography. One such was that these chains as they run through the crystal are so shaped as to repeat the same spatial arrangement of the residues (amino acids) a fixed number of times. Thus a helical chain has a repeat – it returns to the same position as do the steps on a helical staircase – but, it was generally assumed, always after an integral number of steps. Pauling's alpha helix had a repeat of one-and-a-half residues – it was non-integral.

In making this break with tradition, Pauling was influenced by the following considerations. First, his studies of the structure of amino acids and peptides caused him to impose certain restrictions on the possible orientations of the polypeptide chains in proteins. Second, although he used Patterson analysis to guide his interpretation of the data, he did not rely solely or chiefly on this approach. Instead he built molecular models to simulate possible conformations of polypeptide chains. By this trial-and-error method he found he could reduce the number of possible structures. Third, using Fourier methods, he calculated the form of the X-ray diffraction pattern that would be produced by a given molecular structure. This meant that the diffraction pattern, instead of being treated as the product of the crystal was treated as the product of the molecule. In the case of molecules with long chains like DNA, if the molecule were helical, calculation showed that it should produce a diffraction pattern of a particular form – a maltese cross with clear spaces within the cross. Rosalind Franklin produced such a pattern in 1952, and Crick learned about this in 1953 and realised that the DNA structure might not prove to be so difficult to solve.

In 1951, keen to emulate Pauling's success with polypeptides, Watson and Crick built a model of DNA that the London group quickly rejected. Their model had the backbones of the DNA chains on the inside and the four constituent bases projecting outwards. The Cavendish Professor of Physics, Sir Lawrence Bragg, then forbad them

to stray into the territory of the London group again. Time passed during which little news of Rosalind Franklin's progress filtered through to them. Committed to solving the structure by the Patterson method, Franklin was moving cautiously a step at a time. This way Watson and Crick judged that the prize for solving the structure would go to Pauling. Sure enough, in January 1953 a copy of Pauling's proposed structure for DNA came into their hands. Exasperated at their failure to get more action from London, Watson appealed to Bragg to lift the moratorium on him. He was successful. This time he built models with the backbones of the helices on the outside and the bases on the inside. This focused attention on the bases. Fortunately there had been a programme of research at Cambridge University since 1945 on the structure of these bases. Thus Watson pored over the doctoral thesis of June Broomhead, for here were diagrams of bases packed in the crystal, their proximity to one another just about the distance Watson reckoned he needed to fit pairs of them inside the DNA helices and held to one another by hydrogen bonds. Having decided on two helices in the structure, he overcame the problem of keeping the sizes of the pairs of bases constant by pairing a large base – a purine – with a small one – a pyrimidine, adenine with thymine [A–T] and guanine with cytosine [G–C]. There were two further sources that became available at this time – news of Franklin's fine 1952 diffraction pattern of DNA and a copy of Franklin's report on her work to the Biophysics Committee of the Medical Research Council. From the latter Crick learned that the symmetry of the crystalline DNA was what is called 'C2 monoclinic', and he knew that this almost certainly meant that the two chains in the DNA run in opposite directions – they are anti-parallel. Using model building and observing strictly the accepted distances of separation of atoms, they built their famous double helix.

Since the unfortunate early death of Franklin, the award of the Nobel prize to Watson, Crick and Franklin's colleague, Maurice Wilkins, and the unfavourable depiction of Franklin in Watson's book *The Double Helix,* there has been criticism that Franklin has received insufficient credit. Responding to this, Crick has described her work as

first class. 'It is difficult', he added, 'to see how it could be bettered.' But in her professionalism she was not prepared to guess and perhaps get it wrong. Watson and Crick were – even a second time! They were ambitious and competitive, but they sensed the great importance of the structure and did not want Pauling, yet again, to get there first. So they were not put off their quest for a helical model by contrary experimental data – such as non-helical evidence – or functional difficulties – how do the helices unwind from one another in order to duplicate? Their judgment as to which evidence to trust and which to ignore was sound.

They had several advantages over Franklin. Foremost was their very effective collaboration. As Crick has remarked, it is so easy to be led astray in science, but having a collaborator to play devil's advocate provides a check at each turn. The Cambridge unit, too, provided the ideal place for solving a structure like DNA, for nowhere else in the world had so much effort been devoted to the study of the structure of large molecules of biological importance. Much of the 'know-how' gleaned there in studying the proteins could be applied to DNA. Its bases, the purines and pyrimidines, had been studied there by Cochran, Clews, and Broomhead. Alexander Todd, the University's Professor of Chemistry, had established the nature of the sugar-phosphate linkage of DNA's backbone. As a centre of X-ray crystallography, supported by its host the Cavendish Professor, the Cambridge unit was attractive to American researchers wishing to work in Europe. Thus Jerry Donohue, visiting from Pauling's laboratory in Caltech, advised Watson about the structural formulae of the bases, incorrectly represented in the textbooks.

Although structural studies of the proteins had a home in Cambridge, the nucleic acids (in contract to their bases) did not get a home until the expansion of the MRC Unit in 1958. Neither did the genetics of phage – the experimental system that had begun the era of molecular genetics – until Sydney Brenner came to the Unit that year. Links between the university and the Unit might have been forged through teaching, but university departments and their appointing committees did not look favourably upon candidates for university

posts from the MRC Unit. Indeed, once the supportive Bragg had left Cambridge for London, the MRC Unit itself came under threat, for the physics department now needed the space occupied by the Unit. Help came not from biology, biochemistry, or genetics, but from medicine. Moved to the Postgraduate School of Medicine outside the town center, and joined by Fred Sanger's group, the Unit in 1958 became the MRC Laboratory of Molecular Biology.

Turning back to the summer of 1953, Watson and Crick, after a period of concern, became more confident about their structure for DNA. The evidence in its support was increasing. Then came a letter from the physicist, George Gamow, suggesting how the structure of the double helix could act as a template for protein synthesis. This focused the attention of Watson and Crick on the topic, and checking out Gamow's list of the twenty amino acids for which he believed there would be a DNA template, they reconstructed it, producing the twenty that we know today are encoded in the genes. Returning to the United States that autumn as Research Fellow at Caltech, Watson's collaboration with Crick was now only sporadic. But their Cambridge collaboration had set them on careers that have made them two of the most significant scientists of the twentieth century and winning them honours from around the world. But as Crick has remarked: 'Rather than believe that Watson and Crick made the DNA structure . . . the structure made Watson and Crick.' And he added that he was 'almost totally unknown at the time, and Watson was regarded, in most circles, as too bright to be really sound'.

Further reading

Francis H.C. Crick. 1988. *What Mad Pursuit*. London: Weidenfeld & Nicolson.

Francis H.C. Crick. 1994. *The Astonishing Hypothesis. The Scientific Search for the Soul*. New York: Scribners.

Soraya de Chadarevian. 1996. Sequences, conformations, information: Biochemists and molecular biologists in the 1950s. *Journal of the History of Biology*, **29**, 361–86.

Soraya de Chadarevian. Forthcoming. *Molecular Biology at Cambridge: Forming a Discipline*. Cambridge: Cambridge University Press.

Horace Judson. 1996. *The Eighth Day of Creation. Makers of the Revolution in Biology*, expanded edition. Plainview, NY: Cold Spring Harbor Laboratory Press.

Robert Olby. 1972. Francis Crick, DNA, and the central Dogma. In *The Twentieth Century Sciences. Studies in the Biography of Ideas*, ed. Gerald Holton. New York: Norton.

Robert Olby. 1994. *The Path to the Double Helix. The Discovery of DNA*. New York: Dover.

James D. Watson. 1980. *The Double Helix. A Personal Account of the Discovery of the Structure of DNA. A Norton Critical Edition*, ed. Gunther S. Stent. New York: Norton.

James D. Watson. 2000. *A Passion for DNA. Genes, Genomes, and Society.* Cold Spring Harbor: Cold Spring Harbor Laboratory Press.

James (Dewey) Watson (1928–) US molecular biologist; a co-discoverer of the double helical structure of nucleic acids and their place in molecular genetics. Watson's boyhood enthusiasm for bird-watching led him to entry, aged fifteen, to Chicago University where he graduated in zoology when only nineteen, He worked for his Ph.D. at Indiana University, Bloomington, studying phages (bacterial viruses), learning much about bacterial viruses and biochemistry and becoming convinced that the chemistry of genes, then little understood, was of fundamental importance for biology. A fellowship took him to Copenhagen in 1950 to study bacterial metabolism, but soon his enthusiasm for DNA led him to Cambridge and to collaboration with Crick in the Cavendish Laboratory. He shared the Nobel Prize for Physiology in 1962 with Francis Crick and Wilkins.

Francis (Harry Crompton) Crick (1916–) British molecular biologist: co-discoverer with Watson of double-helix structure of DNA.

The outstanding advance in the life sciences in this century has been the creation of a new branch of science: molecular biology. In this, Crick has been a central figure and its key concept, that the self-replicating generic material DNA has the form of double helix with complementary strands, is due to him and J.D. Watson.

Robert Olby began the study of the history of biology at Oxford University where he entered into what was to prove a life-long interest in the history of genetics. Related to this field has been his study of the history behind the discovery of the structure of DNA. He taught the history of science at the University of Leeds for many years. Now, based at the University of Pittsburgh as Research Professor, he is completing a history of biology for the Fontana History of Science Series.

19 Mary Cartwright

TOM KÖRNER

Mary Cartwright never symbolised anything except herself, but her life echoed that of the generation of real and fictional 'new women' who after the First World War took the men on at their own games and trounced them. She was born in 1900 into a middle-class family with a tradition of public service. Her father was rector in a small Northampton village. Two of her brothers were killed in the First World War and one became deputy chairman of the British Steel Corporation.

She hesitated between history and mathematics at Oxford but chose mathematics. The first two years were hard, she found the lecture rooms overcrowded (they were filled with men returning from the war) and she felt herself ill prepared for the course. When she received only a second in the second year examination she seriously considered switching to history. However she persevered, and at the end of her third year an event occurred that was to change the course of her life.

> [A friend] asked me to accompany her to a party on a barge in Eights week. I think the chaperon rules of those days permitted her to accept the invitation provided that she had a woman companion of *any* status and not necessarily a formal chaperon. Actually I do not remember much about the party, but [one of the guests] was V.C. Morton, later professor at Aberystwyth. He was a year ahead of me in work and he advised me to read Whittaker and Watson's *Modern Analysis* and to go to Hardy's Monday evening class. . . . I suppose I owe my career to him, but I do not think I ever spoke to him again. . . . My official teachers never mentioned Hardy.

She read *Modern Analysis* ('as far as I could') over the summer and obtained special permission to attend Hardy's class. Even today, a student who reads 'Whittaker and Watson' by themselves is likely to

FIGURE 19.1 Dame Mary Cartwright (1900–98), by Stanley Spencer, 1958. *Source:* Girton College, Cambridge.

prove outstanding, but the association with Hardy was still more significant. Working separately and in collaboration Hardy and Littlewood dominated British pure mathematics between the wars. (Their collaboration is described in another chapter, but where lives overlap chapters will as well.) With notable but isolated exceptions mathematics in Britain had been seen as a matter of teaching and examining rather than research. Hardy and Littlewood represented the triumph of the reform movement which sought to professionalise British mathematics along the well-established lines of the German and French models. Not only were Hardy and Littlewood world figures in mathematics (it is said that when the leading German mathematician Klein visited Hardy at Trinity he wrote an indignant letter to the

Master saying that since Hardy was the best mathematician in England, he should have the best rooms in Trinity) but they had many research students, ran research seminars, and so on. For the first time there was a genuine school of pure mathematics in Britain.

The tone of the Hardy–Littlewood school was confident, intellectual and progressive. Hardy opened his inaugural lecture as Professor at Oxford by declaring that, 'If I could attain every scientific ambition of my life, the frontiers of the Empire would not be advanced, not even a black man would be blown to pieces, no-one's fortune would be made, and least of all my own.' But though the tone was progressive it was not egalitarian. Those in it were not judged by class, race (Hardy considered the Indian mathematician Ramanujan to be 'his greatest discovery'), or sex, but they were judged.

In 1923, Cartwright received a first class in her final examinations. This possibility had been open to women for only a couple of years and Mary Cartwright was the first woman actually to take finals. She then went into school teaching, but chafed under the rigid control on content and delivery imposed on her. She recalls that when

> we were ploughing through quantities of examples on stocks and shares, a voice at the back of the room said, 'Is brokerage *really* only 1/8th per cent.? I am sure my father earns more than that.'

In 1928 she returned to research, joining Hardy's group of students at Oxford. She rapidly obtained new results and completed a D.Phil supervised by Hardy and Titchmarsh. Littlewood was external examiner and recalled that the first question by the other examiner was so silly as to make Cartwright blush, but that he restored her composure with a wink.

She now moved to Cambridge, to a research fellowship at Girton. The organisation of research still depended on individuals and much was still college based rather than university based.

> When I came to Cambridge in October 1930, Hardy was still at Oxford and [there] was the Littlewood Conversation Class in his

room at 5pm, preceeded by a very good tea for all. It had been going for at least a year, probably since he became Rouse Ball Professor and Littlewood himself *always* did *all* the talking. The following year Hardy returned to Cambridge . I had attended his seminar in Oxford which took place at 8.45 (or rather 9) pm, and was followed by tea and biscuits and conversation about mathematics in general and mathematicians, ending about 11pm. Usually one of his students talked about his own work; Hardy usually talked once during a term and very occasionally distinguished visitors. I remember Landau coming twice and Besicovich, but never Littlewood. The class consisted entirely of Hardy's own research students, or former research students returning for a visit, not more than six of us in all. I asked Hardy if he would be having a seminar and he replied that he would probably come to some arrangement with Littlewood. At Littlewood's first class, Hardy came in late, helped himself liberally to tea and then began to ask questions. I cannot remember what they were, but my impression is that Hardy tried to pin Littlewood down on the details, whereas Littlewood was trying [in his own words] to give the class 'the chance to catch on at once to the momentary point and take the details for granted.' Anyway Littlewood said 'I am not prepared to be heckled' and I do not think they were ever present together at any subsequent class. For a term, or perhaps a year it continued, one week with Littlewood doing all the talking, the next as Hardy's class with a different person talking each fortnight, still in Littlewood's rooms. Littlewood eventually ceased to participate in any way though the class continued to take place in his rooms. Then it moved to the old Combination Room and finally to the Arts School, still at 5pm with tea first, and still in the Lecture List as Professor Littlewood and Professor Hardy: Conversation Class, but it became a much larger affair as Hardy attracted foreign mathematicians and English mathematicians on sabbatical leave to Cambridge. It was a very good meeting ground, but not with the intimate contact with the great man we enjoyed at Oxford.

In spite of this Littlewood and Hardy formed the most successful partnership in mathematical history. Their 'axioms of collaboration' (given in greater detail in the essay on that collaboration) included the following rules:

1. When one wrote to the other, it was completely indifferent whether what they wrote was right or wrong.
2. When one received a letter from the other, he was under no obligation to read it, let alone answer it.

Cartwright attended Littlewood's courses and classes and wrote and talked to him about them. About one topic she wrote that 'Littlewood must have made made me feel that it was an exciting problem and that *he* would be interested in any substantial step towards a solution.' In another course given by Collingwood she learnt of a result called Ahlfors' distortion theorem.

> When I learnt about Ahlfors' distortion theorem. I distinctly felt that *this* was the method to use, but when I looked at [a proof of special case which was already well known] I thought that I could make *that* work.... So I sent my first attempt to Littlewood *not* using Ahlfors' theorem. I got back a postcard with [a drawing of a] twisted snake... showing that my proof contained a common fallacy. Not long afterwards in my bath one night I had a 'great idea'... and I soon sent him a new proof using Ahlfors' theorem....
>
> For several days I heard nothing. Then, when I was punting an elderly Fellow of Girton along the Backs, I saw Littlewood on the path beside the bank and asked him if he had read my manuscript. He said 'No,' and showed considerable reluctance to do so, but I pressed him. He read it and seemed impressed.

Cartwright's theorem, like her thesis and the rest of her work before 1938, belongs to classical complex variable theory. (The lay reader should be warned that the use of the word 'complex' here is a historical accident and bears no relation to normal usage.) This theory was already a century old and extremely well developed when she started

work. Such was its level of sophistication that the pure mathematicians who worked on it would, I think, have been startled by any suggestion that it had any applications. However, it does, and Cartwright's theorem is still frequently quoted and applied in signal processing.

Cartwright's work on complex variable theory brought her into touch with topological methods. (If the reader simply substitutes 'geometrical' for 'topological' the spirit of the matter will be preserved.) It is obvious to the reader that a circle divides the plane into an inside and an outside and mathematicians would agree. It is equally obvious that any 'loop' divides the plane into an inside and an outside, but mathematicians were shocked to find how hard it was to prove this 'topological' theorem. If you stir a cup of coffee then at every moment there is at least one point in the liquid which is exactly where it started. That is, perhaps, much less obvious to the reader, and mathematicians were delighted when this 'fixed point theorem' was obtained from 'topological' considerations. Cartwright remembers that, 'Hardy seemed to me to have almost a phobia about doing analysis which involved thinking geometrically... whereas Littlewood enjoyed it.'

With Hardy and Littlewood's support her Girton fellowship was renewed and in 1935 she became a University assistant lecturer (followed in due course by a university lectureship). In 1936 she also became Director of Mathematical Studies at Girton.

As a Director of Studies myself, let me remark that the post of Director of Studies in mathematics is rarely a sinecure and that the post was particularly onerous at the women's colleges before the 1960s. Unlike the men's colleges which had always contained a fair proportion of 'non-intellectual' students, the women's colleges were under internal and external pressure to keep the standards of all their students high. However, few girls received the kind of education which would encourage or enable them to do mathematics at Cambridge. Many of those who did have the required education were put off by the reputation for difficulty and competitiveness of the Cambridge course. On the other hand, able boys could, if they were lucky enough to attend the right school, obtain an extraordinarily good pre-university training

in mathematics. It was these able and prepared boys who set the tone of the system, and it was the job of the women's Directors of Studies to ensure that their charges would at least survive and preferably flourish in a course not designed for them. Her pupils remembered her with gratitude.

In January 1938 the radio research board issued an appeal to pure mathematicians for help in solving systems of non-linear differential equations arising from the behaviour of thermionic valves. It asked for 'really expert guidance' if only to prevent 'the waste of time and energy spent in pursuit of a will-o' the-wisp' of explicit analytic solutions having something of 'the same comparative simplicity and utility as those available for the linear differential equations' with which the engineer is familiar (the quotations come via Cartwright from the original document). Cartwright describes the equation cited as 'heavy and uninteresting'. We know that the British were rushing to develop radar, it must have been clear that the appeal related to military applications in the probable coming war.

Cartwright read it and noted that '[It] gave four references of which the the most important was "Non-Linear Oscillations" by B. van der Pol, P.I.R.E., Vol 22 (1934). What I asked in some annoyance did P.I.R.E stand for? *Proceedings of the Institute of Radio Engineers*.' She 'worked backward into the subject from the references' and though she speaks highly of van der Pol's paper she notes that, 'Like many authors, myself included, he did not read all the work to which he referred.' During her trawl through the literature she came across several problems which she hoped would interest Littlewood. Some did not but some did, and thus began a ten year collaboration.

A differential equation may be thought of as a mathematical black box with an input (which we may choose) and an output which we want to know. It may be used as a model for a physical system like an electric circuit (in which the output is a current reading) or the firing of a shell from a gun (in which case the output is the position of the shell). The reader probably thinks of the mathematician as solving such a black box problem by writing down a mathematical formula for

the output. For 250 years this is what mathematicians tried to do, but it gradually became clear that, although such a formula could be written down in some important cases (for *linear* systems), most differential equations simply did not have solutions (outputs) which could be written down as mathematical formulae.

(The reader may wonder what a linear system is and why it is relatively easy to handle. Briefly, for such a system the output from the combined sum of two inputs is the sum of the outputs of the two separate signals. This makes it possible to build up the solution to complicated problems from simpler ones.)

Cartwright tells how:

> since then I have discussed non-linear problems of automatic control, vibrations of steel plates, the radiation of stars, movements of planets, homopolar dynamos, guided missiles, wireless masts, flagellar movements of spermatozoa in rats and other physical and biological problems . . . . I have seldom, if ever, produced a solution; sometimes I have been able to suggest a method, and sometimes the discussion has been profitable only in showing that the mathematical problem as stated could not represent the physical situation to which it it was supposed to correspond. . . . [The engineer] tends to ask for 'a solution' rather than saying what he wants to know about the solution, and does not think it worth mentioning the important information about the solutions which is implied by the physical situation.

If we cannot write down a mathematical formula, what can we do? It is well known that the weather prediction 'the weather tomorrow will look like the weather today' is a very good one. In the same way, if we wish to know the price of bread next week our first guess is that it will be the same as today, our second guess will take into account inflation (the rate at which the price of bread is rising), our third guess will be refined by taking the rate of change of the inflation rate, and so on. This is the basis by which differential equations are solved numerically. At that time the only way to do this was by hand and was only

done in cases of extreme necessity; nowadays computers make the task easy. The idea can also be used to obtain mathematical formulae which are valid 'for small periods of time'. Unfortunately both the computational and the mathematical methods based on these ideas break down (for reasons which are analogous to those which prevent us predicting the weather in a year's time or the price of bread in a century's time) if we wish to describe the solution (output) over a long time.

The alternative path, which Cartwright and Littlewood were to take, is to try to describe the nature of the solution. Thus we could say that the output of the black box remains bounded or grows without limit or that the output slowly settles down to a periodic pattern. Consider a long-distance runner running round a track. Because of small variations in speed we can not predict where the runner will be in an hour's time, but we have a very good idea of the nature of path he will be following.

Why were Cartwright and Littlewood pioneers? The first answer is that hard problems in mathematics are two a penny. It is solutions to hard problems which are rare and valuable. The second answer is that with the passing of time new tools like topology had entered the mathematician's armoury. I should like to add a third answer.

The notion of a black box is very general and we must have some way of deciding which black boxes (differential equations) are likely to be interesting. One way is to look at systems that occur in physics or engineering. If nature finds a differential equation interesting, so should we. Moreover when nature provides a differential equation it also provides a solution. (Just look at how the physical system behaves.) It is much easier to tackle a problem if you have clues as to how it may come out. Before the twentieth-century, most natural systems that could be observed were frictional. If you leave such a black box alone it runs down and stops. The only natural systems that do not run down occur in astronomy where we may observe the endless dance of the planets and stars. It was thus natural that Cartwright looked at work by Poincaré, Birkhoff and others dealing with celestial mechanics.

The new technology of radio valves and amplifiers produced black boxes with an internal supply of energy. Such systems behaved neither like earthbound natural systems that ran down nor like celestial systems which did not stop, but could use the internal energy source to exhibit more and more violent behaviour. A typical example is the whistling which used to affect public address systems. To some extent the theory of linear systems could still be used and engineers still exhibit immense ingenuity in getting naturally non-linear systems to behave linearly. However, the mathematical theory of linear systems says, in effect, that they must either run down or explode. Since (at least when working properly) the new systems did neither, mathematicians were now presented with a genuinely new collection of black boxes.

Faced with a new collection of problems, mathematicians can either try to find general methods applicable to all of them or try to make an in-depth study of one of them in the hope that such a detailed approach will ultimately shed light on the entire class. Either approach may be profitable, but it was typical of Littlewood (and of the Hardy–Littlewood school in general) to choose the investigation of a single difficult problem. This was the so-called Van der Pol equation. (Van der Pol, whom I mentioned earlier, was a very mathematically minded engineer who tried to find differential equations which illustrated general problems but which were simple enough to investigate mathematically. When mathematics would carry him no further he set up circuits satisfying the differential equations and investigated the system empirically. Cartwright and Littlewood now sought to explain and confirm the empirical observations.)

The two collaborators contrasted in many ways. Cartwright writes:

> In the early stages of my investigations I read bits of many papers and books, often picked up at random from library or bookshop shelves, in the hope of obtaining some crumbs of mathematical information to suggest a line of mathematical attack; my understanding of the technical side was very superficial.

On the other hand, Littlewood felt that

> one should not pay too much attention to the existing literature on a problem. For if previous writers had failed to solve it, it was probably because they had tried a wrong method. He gave no obvious sign of having read any of the literature from which I extracted our problems and his methods were quite different from those of the radio engineers.

Unlike Cartwright, Littlewood already knew the literature of classical mechanics. He had a further source of insight from his military service in the First World War, which he spent calculating shell trajectories for anti-aircraft fire. Hardy, with his pacifist sympathies, wrote the mathematics of war '(whatever one may think of its purposes) is repulsively ugly and invariably dull; even Littlewood could not make ballistics respectable, and if he could not, who can?' Perhaps Littlewood's work on differential equations may be said to have transmuted lead to gold. Certainly he seems to have thought in dynamical terms.

Having said this, it seems worth noting that too deep an immersion in one set of engineering or physical problems may blind one to connections with other problems. To quote Cartwright:

> it seems difficult for [engineers] to think in abstract mathematical terms, the symbols to them seem to mean the engineering concepts, currents and circuit constants such as impedance and inductance. This is important in two ways. The engineers have mental reservations and can check at every stage because they visualise how the physical system works. On the other hand they find it difficult to apply the mathematical process used in one field to any other physical problem, even if they are just as relevant there.

Cartwright remembers that:

> As regards the techniques of collaboration, we did it all by letters with occasional short discussions of particular points, never at a blackboard, never in my room, seldom if ever in his.

Littlewood described her as 'the only woman in my life to whom I have written twice in one day'. They agreed that 'in a true collaboration the authors do not know *who* had which idea first' and published some papers jointly and some under a single name. Like others of his collaborators Cartwright found it difficult to persuade Littlewood to complete a paper. She says that 'his mind was so full of ideas that he found it difficult to stop', but there may have been a psychological element as well. Littlewood suffered from periods of mental depression. These were cured in the mid 1950s, but were at their most severe during the years covering his collaboration with Cartwright. Nothing was published until 1945, but 'he made a special effort to complete an announcement of some of our results . . . to help me as a candidate for the Royal Society'. (The Royal Society had just been opened to women. Cartwright was among the first elected and the only female mathematician elected until Dusa MacDuff.)

The results obtained by Littlewood and Cartwright looked extraordinary to anyone whose intuition was based on linear systems. Under certain conditions the Van der Pol system could settle down into one of two different periodic oscillations. (It is as if striking a bell could give two entirely different sounds.) Moreover the way in which the system settled down to one or the other state was extraordinarily complex. Littlewood wrote:

> Two rats fell into a can of milk. After swimming for a time one of them realised his hopeless fate and drowned. The other persisted, and at last the milk turned to butter and he could get out. Miss Cartwright and I . . . went on and on at the thing with no earthly prospect of 'results': suddenly the dramatic fine structure of solutions stared us in the face.

They had started by attacking the most interesting problems 'and that', as Cartwright said to Littlewood, 'meant the most difficult', leaving the duller parts to later. But 'when we came to the basic duller parts they were not so easy as he had thought'. As Littlewood wrote to Cartwright, 'All details have a nasty way of ramifying into difficulties.'

The resolution of these difficulties gave rise to more interesting papers. In the course of their investigations they were led to a fixed point theorem which was much more general than the 'coffee cup' fixed point theorem I quoted earlier.

The reader may wonder about the original engineering problem. As might be expected, 'the radio engineers found that, owing to variations in the valves themselves, trial and error was more effective than any mathematical treatment could be'.

In 1948 Cartwright's colleagues at Girton elected her Mistress (head) of Girton. The women's colleges had just been given the same status and rights as the men's colleges, but the change in status did not imply an increase in wealth and they were among the poorer members of the community of colleges. Many people (particularly those who have the position) compare the job of head of college to that of herding cats, but Cartwright did it quietly and effectively. Her duties, as she saw them, included interviewing all applicants for entry to the college, which, apparently, she did in batches of five, lumped together quite irrespective of the subjects for which they applied.

Inevitably her mathematical output diminished, though she still published papers (her best-known work in this period being a collaboration with Collingwood on a very delicate problem in complex variable). However, the Hardy–Littlewood school held that a mathematician's best work is done before fifty and that after that age they should be free to do other things. In 1959 she was promoted to Reader at Cambridge. On her retirement she was made Dame of the British Empire and was thenceforth known throughout Cambridge as 'Dame Mary'. In her retirement she took up visiting professorships at several universities. She helped edit Hardy's papers and in 1980 published two long and very interesting accounts of the Hardy–Littlewood correspondence. She was quietly amused to discover that Axiom 1 (when one wrote to the other, it was completely indifferent whether what they wrote was right or wrong) had been liberally used, and that the correspondence of the two most successful mathematical collaborators in history was sprinkled with elementary errors.

Although she was apparently an excellent and conscientious research supervisor, she only took on a limited number of research students during her career. However this limited number included Walter Hayman, the leading British complex analyst of his generation.

It is natural to ask if she encountered prejudice towards women in her career. She believed that she had not. She was never made Professor at Cambridge, but the same was true for male colleagues of similar standing. (The present galloping inflation in academic rank had not started, and the Cambridge mathematics faculty fondly believed that its professorships could always be filled with a Dirac, a Hardy, or a Littlewood.) Apart from this, she obtained all the honours (the Sylvester Medal, the presidency of the London Mathematical Society, the De Morgan Medal . . .) which a world-class British mathematician might expect.

Possibly her path was made easier by the fact that she shared so many of the values of her male colleagues. (Some values she did not share. 'I was elected to committees on which I was the only woman. I was horrified by the lack of formality and inattention to business of the men.') Certainly her path was made easier by her native toughness and sense of humour.

Littlewood observed that 'One said: "Perhaps I may be stupid but . . ." [and] she would reply: "Well yes, I think so."' One of her strengths was that she saw things as they were, whilst Hardy wanted them in the neat shape of an epigram. Hardy professed the ruthless pursuit of excellence but Cartwright remarked to a friend

> that Hardy could not really think like that because, if he did, he would not take so much trouble over weak students (as we both knew that he did). She replied 'That is not what he thinks but what he likes to think he thinks.'

According to Littlewood

> She is a fundamentally very modest (though dutifully asserting her rights as a representative Academic woman). In earlier days she used to make in the blandest tones the most devastating remarks, almost

all of which landed on Hardy. . . . When there was a question of a mathematical dinner and I suggested that Miss Cartwright should sit next to Hardy, he said: 'Well: but her fast ball is so very devastating.'

At the same time as Cartwright and Littlewood were attacking their problem, Lefschetz, Levinson, and others in the United States, 'also impelled to a large extent by applications connected with the war' were investigating similar problems. Instead of concentrating on one problem, as Cartwright and Littlewood had done, they sought features common to large groups of problems. They drew attention to earlier work of the Russians Kryloff and Bogluiboff. With the publication of the American results at the end of the war, these various streams of thought began to merge.

In some ways the sheer mathematical power Cartwright and Littlewood brought to bear on their problems carried them too far ahead. Mathematicians sometimes refer to 'inventing the carburettor in the stone age' and the bow which Cartwright and Littlewood had strung could not be used by anyone else. Mathematicians were of course delighted with the 'first published examples of a wholly rigorous treatment of a problem in large parameter theory' but few would have the courage to read them through and still fewer to try and use their methods.

However, the ideas of the their papers soon entered the general domain. We now recognise the 'fine structure' discovered by Cartwright and Littlewood as the first example of a 'stable chaotic system'. The underlying general mechanism was elucidated in 1965 by Smale with his 'horseshoe' construction and forms the basis for modern 'chaos theory'.

The work of Cartwright, Littlewood, and their American counterparts changed the way we think about differential equations for ever. They changed the questions that we ask and the answers we expect. In emphasising the topological aspect of the subject they both joined and accelerated a swing now visible over much of mathematics towards a 'regeometricisation' of the way we think of things.

To a pure mathematician such achievements are enough, but the reader may wonder if the engineers who inspired the original investigation received anything back. That publishers benefited is clear from the shelf upon shelf of books on chaos and catastrophe theory, but the general public draws from these books (or at least their titles) a picture of mathematical impotence. Such a picture is mistaken. These theories tell us not only what we cannot predict and control but also what we can. We may not be able to predict the weather arbitrarily far into the future but we can now tell how far into the future our forecasts are likely to be accurate. Chaotic does not mean uncontrollable; the space probe missions to the outer planets are possible precisely because the path of a probe is in some sense chaotic and we can use this fact to give the probe a 'free ride'.

The interplay between mathematics and engineering, like that between mathematics and physics, is a subtle one but one from which both sides gain. From the engineers, mathematicians received new and interesting questions. In return the engineers received a general and suggestive language applicable not merely to radio engineering but to control theory, weather forecasting, space travel, and much else, which they have been quick to exploit.

Towards the end of his book *A Mathematician's Apology*, Hardy reflects on his own life. Two passages from his reflections seem appropriate to conclude this essay. In the first (which has such resonance for mathematicians that both mathematical essays in this book use it) he says:

> I still say to myself when I am depressed and and forced to listen to pompous and tiresome people, 'Well, I have done one thing *you* could never have done, and that is to collaborate with both Littlewood and Ramanujan on something like equal terms'.

And he ends:

> The case for my life, then, or for that of anyone else who has been a mathematician in the same sense in which I have been one, is this:

that I have added something to knowledge, and helped others to add more; and these somethings have a value which differs in degree only, and not in kind, from that of the great mathematicians, or of any of the other artists, great or small, who have left some kind of memorial behind them.

Further reading

S.L. McMurran and J.J. Tatersall. 1996. The Mathematical Collaboration of M.L. Cartwright and J.E. Littlewood. *American Mathematical Monthly*, **103**, 833–45.

The forthcoming memoir of Mary Cartwright in the *Biographical Memoirs of Fellows of the Royal Society* by Professor Walter Hayman should also prove a valuable reference.

Mary Lucy Cartwright (1900–98) was the first woman to take the mathematics finals at Oxford, and obtained a first. After a brief period as a schoolmistress she returned to Oxford where she obtained a D.Phil. In 1930 she went to Cambridge as a research fellow at Girton. Here she remained, becoming University Lecturer (1935) and Reader (1959) as well as Mistress of Girton (1949–68). She was made a Dame of the British Empire in 1969 and was the first female mathematician elected to the Royal Society (1947).

She was one of the foremost British mathematicians of her time, working in the fields of complex variable and differential equations. Her most lasting impact derives from her joint pioneering work with J.E. Littlewood on differential equations.

T.W. Körner is a University Reader at Cambridge and Director of Studies at Trinity Hall. He works in the field of Fourier analysis and has published two textbooks on this subject as well as a semi-popular text *The Pleasures of Counting*.

20 Joseph Needham

GREGORY BLUE

After winning acclaim for his monumental work on the history of Chinese science, Joseph Needham was occasionally introduced to fellow scientists who expressed appreciation for what they took to be his father's pioneering research on chemical embryology. The two women central to his adult life – his first wife Dorothy, and his long-time collaborator and eventual second wife Lu Gwei-Djen – used to enjoy recalling the surprise on the face of new acquaintances when they realised the biochemist and the sinologist were the same man. Like A.N. Whitehead, whose philosophical views and breadth of outlook he absorbed as a young man, Needham made the history of science his priority only after first gaining a reputation as an innovative scientific researcher. His later orientation was nevertheless solidly grounded in sensibilities he cultivated from youth.

Joseph Needham was born in London in December 1900. Looking back on his life in the 1970s, he accounted for the distinctive features of his character in terms of the influence of his parents. His tendency to embark on expansive projects he saw as reflecting the artistic temperament of his Irish songwriter mother, while his scientific propensities and broad religious interests he ascribed to his English father, a physician of Anglo-Catholic conviction and Gallophile tastes. Like George Bernard Shaw, one of his youthful culture heroes, Needham later explained his creativity, and particularly his abiding desire to build bridges between various areas of interest, as the consequence of a desire to reconcile parents whose personalities and opinions often clashed. Whatever its origins, this commitment to bridge building pervaded his career. At Oundle School, he acquired several traits that would remain with him for life – a decidedly evolutionary world-view, a solid commitment to socialism and a fondness for practical benchwork. Having

chosen to read medicine at Cambridge, he applied for admission to Gonville and Caius College – the college of William Harvey, he often later recalled. Once admitted, he remained a member of Caius until his death in 1995. Though his outspoken socialism made him unpopular with many older fellows between the wars, this highly traditional college served as a central point of institutional reference for him throughout his career, his devotion to its rituals and formalities often puzzling fellow socialists almost as much as his lifelong high-church Anglicanism. After completing undergraduate and graduate studies, he was elected a Fellow of Caius in 1924. He became President of the Fellows in 1959 before serving two terms as Master from 1966 to 1976. Only in the 1990s did he finally step down as Senior Fellow when failing health prevented him from dining in hall.

As an undergraduate his fascination with biochemistry was sparked by the lectures of Frederick Gowland Hopkins, the founder of British biochemistry, and Rudolph Peters, later Professor of Physiology at Oxford. After his father's death in 1920, Needham looked to Hopkins as a surrogate father-figure. Following a year of post-graduate laboratory work in Germany, he returned to Cambridge to begin graduate studies in the recently established Department of Biochemistry. In 1924, he was awarded his Ph.D. and the Caius fellowship that allowed him to begin his research career. That same year he married Dorothy Moyle, another outstanding biochemist, four years his senior and of Quaker background. Playfully thumbing their noses at convention, the young couple held their wedding on Friday the 13th. In the spirit of sexual openness then fashionable they committed themselves to a 'modern' marriage, an arrangement that did not stop them from spending the next sixty-three years together in loving union. With its exhilarating blend of cutting-edge science, left-wing politics, and lively mixed company, the Dunn Institute of Biochemistry provided them with a congenial professional home between the wars, with Joseph rising to 'second in command' by the mid 1930s.

As a new field, biochemistry in the twenties met with resistance from practitioners of the older, established disciplines. Many biologists

dismissed the idea that chemistry could contribute anything useful to the understanding of life, while among physicists and chemists there was strong suspicion that researchers who occupied themselves with living organisms would never attain the standards of experimental and theoretical precision expected of rigorous scientific work. Known as one of the bolder researchers at the Dunn, Needham focused his research on experimental embryology and in 1923 began the project that would occupy him for the next decade and first establish his international reputation. This study, which grew out of his doctoral dissertation, focused on identifying the chemical processes by which the fertilised ovum developed into a fully formed organism. Concentrating on avian and amphibian eggs, he was soon generating experimental results that underlined the fecundity of biochemical methods for elucidating processes of obvious relevance to the constitution of life itself. Among his more striking observations at this stage was the insight that over the course of its development the embryo draws its energy from a succession of different sources, relying first on carbohydrates, then on proteins and finally on fats. Mixing work at the Dunn with regular trips to major marine biology stations, he pursued his research with a combination of innovative experimentation and theoretical aplomb grounded in vast knowledge of the scientific and philosophical literature. He was appointed Demonstrator in Biochemistry in 1928 and William Dunn Reader in Biochemistry in 1934, in the meantime publishing his first *magnum opus*, the three-volume *Chemical Embryology* (1931), which paved the way for his election to the Royal Society in 1941.

A prominent public spokesman for British biochemistry between 1925 and 1940, Needham presented the new field to the broader scientific community and the educated public alike, while at the same time emerging as a familiar name among the younger generation of general science writers. In these capacities, he published a steady stream of works on the history and philosophy of science. In *The Sceptical Biologist* (1929) and *The Great Amphibian: Lectures on the Place of Religion in a World Dominated by Science* (1932), and eventually in

Time, The Refreshing River (1943) and *History is on Our Side: A Contribution to Political Religion and Scientific Faith* (1946), he addressed issues related to the place of science in society and the relations of science with other dimensions of social life, particularly religion and politics. Among the notable intellectual sources for his personal world-view were Whitehead's process philosophy with its notion of distinct levels of organisation, C. Lloyd Morgan's theory of emergent evolution with its stress on nature's capacity to generate new qualities at each level, R.G. Collingwood's theory of contrasting forms of experience, W.E. Barnes' and Rudolph Otto's ecumenical approach to religious faith, and Karl Marx's analysis of social progress as developing through a succession of modes of production, each featuring characteristic technologies and social relations. The broad, differentiated world-view which he articulated in dialogue with such sources, though deemed paradoxical by some commentators and contradictory by others, underlay all Needham's subsequent work. In his view science was one of several distinct types of human experience, along with art, religion, philosophy, and history. Each type was valid in its domain, and all were amenable in principle to the others, but none was capable of constituting absolute truth; each was able to check the claims of the others to absoluteness, science being particularly important in this regard as an antidote to the claims of religion to embody eternal truth. Yet, although he adhered to this fundamental vision for the rest of his life, his thinking on certain aspects of it evolved over the years.

In the early 1930s, under the influence of J.H. Woodger, he abandoned his earlier neo-mechanist view of the nature of science and adopted an organicist interpretation that highlighted possibilities for examining diverse natural mechanisms within a holistic analytical framework. In 1932 he and Woodger formed the Theoretical Biology Club, a discussion group founded in the hope that an interdisciplinary team might prove an effective medium for fostering a theoretical revolution in the biological sciences. Reflecting the group's early deliberations, Needham's *Order and Life* (1936) outlined a plan of research

geared to developing a multi-dimensional biochemistry and a dynamic morphology which together would allow for analysis of the entire range of organising relations from the molecular and cellular levels all the way up to the levels of the organ and the independent organism. This vision underlay his Rockefeller-funded collaboration with the embryologist Conrad Waddington aimed at discovering the agent responsible for inducing embryological differentiation. Though that project ultimately showed induction to be a far more complex process than expected, the work carried out by the Needham–Waddington team pushed research forward in several important areas. Needham's systematic presentation of the programme's results in *Biochemistry and Morphogenesis* (1942) included his model of the three-dimensional structure of the cell, or cyto-skeleton, a valuable theoretical innovation. However, his proposal for an interdisciplinary institute in which experts in diverse fields would collaborate on devising new approaches to fundamental problems in the life sciences ran into insuperable objections in the mid-1930s from senior authorities who deemed his historical and philosophical interests incompatible with first-rate scientific work.

Needham's passion for history expressed itself from his student days in a consistent concern with placing scientific problems in historical context. During the 1920s he and Dorothy were befriended by two of Britain's leading historians of science, Charles and Dorothea Singer, who opened their home and their library to them. The scholarly fruits of that friendship included nearly 200 pages on historical subjects in *Chemical Embryology*. An expanded version of that exposition appeared as *A History of Embryology* in 1934. By that time Needham had begun shifting to the Marxist view that scientific and technological change should be understood not just in terms of the appearance of new theories and experiments, but also in relation to the social, economic, and political forces that shaped the aims and outlooks of scientists in a given society. This approach, which provided a framework for bridging the divide between social history and the history of science, was reflected in two essays published in the collaborative *Christianity*

and the Social Revolution (1935): one on the attitudes of early Royal Society members towards the English Revolution; the other on the historical relations between science, religion, and socialism. Not long thereafter, Needham obtained permission from the Biology Faculty to organise Cambridge University's first series of regular lectures on the history of science. Working with an interdisciplinary committee of scientists and historians, he inaugurated the series in 1936 and continued to organise it until the war. He and Walter Pagel published the first year's papers in C.P. Snow's Library of Modern Science in a volume entitled *Background to Modern Science* (1938).

Needham's emphasis on placing the history of science in the context of social history parallelled his insistence on the social and political responsibilities of scientists and on making the public good a central consideration in the formulation of research priorities. A Labour Party stalwart, he played a central role in re-vivifying the Association of Scientific Workers in the late 1920s and served as its representative on the Cambridge Trades Council throughout the Depression. His knowledge of Germany and friendship with German scientists made him a staunch opponent of Nazism: within months of Hitler's taking power, Needham was chairing anti-fascist meetings and lending a hand at organising support for exiled scientists and other victims of Nazi persecution. During the Spanish Civil War, he served as national secretary of the Cornford–McLaurin committee, which raised funds for the International Brigades. An outspoken critic of colonialism and the appeasement of fascism as well as of 1930s Conservative policies on public health and civil defence, he shared special concern over the under-funding of science with other members of Solly Zuckermann's 'Tots and Quots', a group whose energetic lobbying was crucial in getting Britain's scientific resources organised for the war. Co-opted onto the Biology War Committee in late 1939, Needham then toured North American universities presenting his paper 'The Nazi Assault on International Science', which helped mobilise scientific opinion for the Allied cause.

On returning to Britain, he became a prominent spokesman for

the Free China Campaign, a role that gave public expression to a fascination with China that dated back to 1937, when three Chinese researchers joined the Dunn Institute. The one newcomer who would have by far the greatest impact on the Needhams' lives was Lu Gwei-Djen, who took up her British Council studentship in Cambridge in the hope that Joseph's support for the Spanish Republic might incline him to similar sympathy with China in its struggle against Japanese militarism. While she completed her doctoral work on nutrition with Dorothy, he immersed himself in Chinese language and history, studying first with her, then with Gustav Haloun, an exile from Hitler's Germany, who had recently taken the chair of Chinese in Cambridge. By the time Gwei-Djen left England for the States on the eve of war in Europe, she and Joseph had initiated a passionate affair which continued off and on for the next eight years. Her return to Europe in late 1947 enabled them to meet again on a regular basis, and their marriage in 1989 finally formalised the relationship after fifty years.

After Japan attacked the Western allies in December 1941, Needham accepted a British Council invitation to undertake a morale-raising visit to China as the representative of the Royal Society. Impressed by his knowledge of their country and by his eagerness to converse in Chinese, his hosts in Chongqing quickly made him aware of their crying need for up-to-date equipment and communication with the broader scientific world. In response, Needham proposed the establishment of a British Scientific Mission to China which would mirror the existing missions in Washington and Moscow. Appointed its head, he was attached to the British embassy as Scientific Councillor and then proceeded to organise the Sino-British Scientific Co-operation Office, an inter-governmental agency. His success in that enterprise and the obvious appreciation of the Chinese scientific community induced him in 1943 to propose the establishment of a global post-war network of scientific liaison offices. This idea coincided with the initiative of the Council of Allied Ministers of Education to establish a post-war United Nations body to coordinate educational and cultural reconstruction. After Needham had lobbied vigorously for the

new organisation to include science, public anxieties over the atomic bomb finally led Allied leaders to give science an equal place with education and culture when UNESCO was initially constituted in November 1945. Its first Director General, Julian Huxley, urged Needham to serve as founding head of the science section. Though nervous about staying away longer from research, Needham judged the opportunities for shaping post-war science in an internationalist direction too important to pass up. Unfortunately, both superpowers perceived UNESCO to be a potential vehicle for espionage by the other. From the outset the new organisation was caught up in the politics of the Cold War. Between April 1946 and March 1948, Huxley and Needham tried to keep the fledgling organisation on a neutralist course as the latter put into place what he saw as the two pillars of the science section: a network of Scientific Field Offices similar to his earlier operation in China and a system of a grants-in-aid to international scientific unions. Increasingly stymied by administrative hurdles imposed by the Truman administration, however, and disappointed at Washington's vetoing of his proposals for UNESCO support to left-wing unions, by mid 1947 he began looking forward to a return to scholarly life in Cambridge. Responsibilities for the proposed UNESCO *General History of the Scientific and Cultural Development of Mankind*, which he had initially planned with Lucien Febvre, were placed in other hands after Huxley left office, though the idea of science as an 'ecumenical' enterprise which Needham devised in that context lived on in his later work.

Back in Cambridge, he resumed his teaching duties in biochemistry, but, as far as research was concerned, his heart was set on the historical project he and Lu Gwei-Djen had first envisioned in 1938. During his wartime travels he had everywhere made a point of investigating traditional technologies and holding discussions on the history of Chinese science, history, and religion, as well as on the current state of Chinese science, and he returned to the West with a large amount of relevant documentation. Any idea of a formal position in history of science was blocked, however, by hostility in key quarters to his vision

of ecumenical science as a foundation for international peace and cooperation. The early death of Haloun likewise left him temporarily without an anchor in the Faculty of Oriental Studies. His alienation reached new depths after he visited China during the Korean War to participate in the International Scientific Commission convened to investigate Communist claims that American forces had used biological weapons in North Korea and Manchuria. Struck by similarities with his wartime evidence of Japanese germ warfare experiments, he publicly endorsed the Commission's conclusion condemning the Americans. Amidst calls for his dismissal from his university and college positions, he settled down in his room in Caius to write what he was convinced would be a very good book.

Science and Civilisation in China (SCC) effectively changed the way the world understood the development of science. In 1971 it won him election as a Fellow of the British Academy, the Royal's Society counterpart for scholarship in the humanities and social sciences, thus making him the first person ever to be elected to both on the basis of separate bodies of work. The first introductory volume was published in 1954 on the understanding that politics would be kept out of the series; by the time the second volume – on science, religion, and philosophy – appeared in 1956, the series was heralded as a monument of scholarship. Not surprisingly, it was not without its critics. Especially in America, the use of Marxist categories led to a stream of condemnations. In Cambridge the sinologist Arthur Waley reported sardonic laughter in one Senior Combination Room after someone announced wryly that Needham was claiming the Chinese had 'invented science'. What he actually argued was that the long-term development of science and technology was an international process, that its record featured contributions by societies around the world and a complex history of fruitful interchanges among them, and that many innovations made in the Chinese culture-area before 1500 had been of crucial importance for the growth of world science.

A decade and a half of research in the Chinese sources together with his work on the history of modern science put Needham in a

position to formulate his case with subtlety, and support it with a mass of new documentation and material evidence. As in his scientific works, much of the conceptual strength of his work on Chinese science derived from an astonishing mastery of diverse bodies of literature, many previously untapped and even unsuspected by other historians of science. In the first instance this included a wide array of Chinese historical sources, from specialised technical and scientific treatises, to official and unofficial histories and encyclopaedias, to various kinds of imaginative literature. He also drew on a solid corpus of secondary studies by modern Chinese writers and Western 'China-hands' who had written on individual scientific or technical topics for purposes of their own. Tapping into the wealth of material in these diverse works, Needham and his collaborators brought forth a mass of evidence regarding technologies and natural knowledge, and assessed the significance of this evidence both in relation to the evolution of Chinese society and in comparison with technologies and ideas attested for other parts of the world. Needham's systematic linkage of science and technology implied a rejection of the cliché that Asian civilisations were spiritual but impractical, but also reflected his conviction that science and technology necessarily conditioned one another in their historical development. The organisation of the SCC series according to modern disciplinary demarcations was in accord with his notion of the universality of modern science; while rejected as anachronistic by anti-positivist critics, this format at least succeeded in translating new and difficult materials into a medium contemporary readers could recognise.

One feature of the project that was not new, though it later came to be known as 'the Needham problem', was the question of why modern science had arisen in Europe, not China. Earlier writers had addressed this issue by highlighting the backward state of Chinese science over the previous two centuries and explaining it as a result of a racial or cultural incapacity for science – a response reminiscent of Molière's character who diagnosed muteness as the consequence of an inability to speak. Rather than beginning with assumptions of Chinese

incompetence, Needham approached the historical record with his experience of his Chinese colleagues' scientific proficiency. From this perspective he proceeded to uncover a vast, complex record of discovery and invention. Knowledge of the Western historical literature led him both to emphasise the comparative richness of pre-modern Chinese achievements and to maintain that a qualitatively superior form of science had emerged in seventeenth-century Europe. Contrary to conventional views that knowledge of nature in non-Western cultures was unscientific because it lacked theory, he maintained that traditional knowledge had been pervaded in the East and the West alike by theories of a particularist type (such as the *yin/yang* dyad and the Galenic four humours), which were closely tied to the cultures that gave rise to them. On this basis he argued that the key historical divide was not between East and West, but between the diverse culture-bound 'traditional' sciences and the distinctively 'modern' form of science, the latter having a special ecumenical potential because its hypotheses were formulated in the language of mathematics, which was particularly amenable to cross-cultural communication. Interpreting a 'traditional' separation of mathematics and nature-lore as reflective of a typical pre-capitalist social divide between manual labour and academic activity, he depicted the modern integration of mathematics and natural knowledge as closely bound up with the early modern expansion of capitalism, which brought together craftsmen and scholars in common pursuit of economic advantage. His answer to the question of why imperial China had generated neither capitalism nor modern science was that the centralised bureaucratic order with its organicist world-view could encourage traditional scientific and technological innovation, but its distrust of merchants precluded modern forms of development. Though later criticised as simplistic, as a heuristic tool his distinction between modern and traditional forms served to break down the assumption that non-Western societies had no scientific and technological achievements worthy of consideration.

In 1959 the third volume of SCC, on mathematics, astronomy, and the earth sciences, appeared as a single book; but the fourth volume,

on physics and engineering, required three separate tomes. Eight books have already been published in the fifth volume on 'chemistry and chemical technology', which treats the alchemical tradition, mining and metallurgy, gunpowder and military technology generally, textile technologies, and paper and printing. The sixth volume, which should eventually include at least six books, examines botany and zoology, as well as the medical and pharmacological traditions. The final, seventh volume considers social and ideological factors that shaped the traditional sciences and technologies in ancient and imperial China, a subject also treated in essays collected in Needham's widely circulated *The Grand Titration* (1969). Throughout successive volumes of the SCC project, he sought to document not only what he called the grandeurs and weaknesses of traditional Chinese science and technology, but also his conviction that many technological innovations made in pre-modern China – including printing, gunpowder, the production of cast iron – spread beyond China's borders to play a revolutionary role in the growth of modern science and society elsewhere. These diffusionist claims have been criticised along with his treatment of Chinese society as a form of bureaucratic feudalism, and his view that Daoism was especially favourable to the investigation of nature. His fundamental case that China possessed an impressive record of scientific and technological achievement before the sixteenth century had nevertheless gained general acceptance among scholars by the 1970s.

As Needham repeatedly stressed, *Science and Civilisation in China* was a collaborative, cross-cultural enterprise from the start. He was the main author of all volumes published until 1980, but much of the legwork in the primary sources was carried out by Wang Ling, who accompanied him to Cambridge in 1946 and worked there until 1957. Shortly before Wang left for a chair in Canberra, Lu Gwei-Djen rejoined the Needhams in Cambridge after retiring from her UNESCO post in Paris. Throughout her years there she had collected materials for SCC and commented on successive drafts. After Wang's departure, she became Needham's chief collaborator, a position she retained until her death in 1991. Over the intervening decades they engaged a growing

number of other collaborators to write on subjects they would be unable to address themselves. On Needham's retirement as Master of Caius in 1976, the two of them amalgamated their libraries to form the East Asian History of Science Library, the collection that now constitutes the heart of the Needham Research Institute, where the production of the remaining volumes of the SCC series is coordinated. By his death in 1995 seventeen volumes had appeared, a monument to scholarship by one of the century's outstanding scientific internationalists and one of its most dedicated believers in the importance of bridging the 'two cultures'.

Books by Joseph Needham

Chemical Embryology. 3 vols. Cambridge: Cambridge University Press, 1931; New York, Hafner, 1963.

Order and Life. New Haven: Yale University Press, 1936; Cambridge MA: MIT Press, 1968.

Biochemistry and Morphogenesis. Cambridge: Cambridge University Press, 1942 (2nd edition, 1966, with new introduction).

Time, The Refreshing River. London: Allen & Unwin, 1943; Nottingham: Spokesman, 1989.

Science Outpost (with D.M. Needham), London: Pilot Press, 1954 [Publication in Chinese, 1947].

Science and International Relations. Oxford: Blackwell, 1949.

Science and Civilisation in China. 7 volumes in 21 parts to date (ongoing, with diverse collaborators). Cambridge: Cambridge University Press, 1954– .

The Grand Titration. Science and Society East and West. London: Allen & Unwin, 1969; University of Toronto Press, 1979.

Three Masks of the Tao. London: Teilhard Centre, 1979.

Celestial Lancets. A History and Rationale of Acupuncture and Moxibustion (with Lu Gwei-Djen). Cambridge: Cambridge University Press, 1980.

Further reading

Maurice Goldsmith. 1995. *Joseph Needham. 20th-Century Renaissance Man*. Paris: UNESCO.

S.K. Mukherjee and A. Ghosh (eds). 1997. *The Life and Works of Joseph Needham*. Calcutta, Asiatic Society.

Noel Joseph Terence Montgomery Needham. Born 9 December 1900 in London; died 24 March 1995 in Cambridge. Married to Dorothy Moyle (1924–1987) and to Lu Gwei-Djen (1989–1991), with no children from either marriage. Educated at Dulwich College Preparatory School, Oundle School, and the University of Cambridge (BA 1922; Ph.D. 1924). Fellow of Gonville and Caius College, Cambridge, 1924–1995; Master, 1966–1976. University Demonstrator in Biochemistry, 1928–1934; Sir William Dunn Reader in Biochemistry, 1934–1966. Fellow of the Royal Society, 1941, Fellow of the British Academy, 1971; Honorary Fellow of the Royal College of Physicians, 1984. Director of the Sino-British Scientific Cooperation Office (Chongqing), 1943–1946; Founding Head of the Science Section of UNESCO, 1946–1948; Director of the East Asian History of Science of Library and the Needham Research Institute (Cambridge), 1976–1990. President of the International Union of the History of Science, 1972–1975. Companion of Honour, 1992.

Gregory Blue is a member of the Department of History at the University of Victoria in British Columbia, Canada. His research concerns intellectual history, Sino-Western relations and the history of science, and his recent publications include *China and Historical Capitalism: Genealogies of Sinological Knowledge* (Cambridge University Press, 1999), edited with Timothy Brook. From 1977 until 1990 he worked with Joseph Needham as a research associate at the East Asian History of Science Library in Cambridge.

21 Molecular biology in Cambridge
M.F. PERUTZ

In 1936, I left my home town, Vienna, for Cambridge, to seek the Great Sage. He was an Irish Catholic converted to Communism, a mineralogist who had turned to X-ray crystallography: J.D. Bernal. I asked the Great Sage: 'How can I solve the secret of Life?' and he replied: 'The secret of life is in the structure of proteins, and there is only one way of solving it and that is X-ray crystallography.' So I became an X-ray crystallographer. We called him the Sage because he knew everything from history to physics. His conversation was the most fascinating of anyone I have ever come across. Actually, what had attracted me to Cambridge was not the Sage. It was the lectures of a young organic chemist in Vienna who told me of the work being done in the biochemistry laboratory headed by Gowland Hopkins, one of the founders of biochemistry.

Hopkins had shown that all chemical reactions in living cells are speeded up by enzymes. They are *catalysed*, chemists say. And he showed that all enzymes are proteins. The remarkable thing in the living cell is that chemical reactions go on at room temperature, in water, at near neutral pH. When chemists make these reactions happen, they need strong solvents or high pressures, or a vacuum, or strong acids and alkalis. In the living cell they take place without any of these, because there is a special protein that speeds up each particular reaction – and speeds it up by a fantastic amount. For instance, hydrogenperoxide is stable at room temperature, but there is an enzyme which decomposes it at the rate of half a million molecules per second. So, the great question was: how do these proteins work? We had no

This chapter (amended here) first appeared in *Cambridge Minds* (ed.) Richard Mason, Cambridge: Cambridge University Press, 1994.

idea. Proteins were black boxes. All we knew about them was their chemical composition.

Genes were also believed to be made of protein. There was another great man at Cambridge: J.B.S. Haldane, one of the most imaginative scientists of this century. He had shown that enzymes are controlled by genes, but he believed that genes – which, we already knew, lie in chromosomes – are also made of protein. This is why the Sage said that the secret of life lies in the structure of proteins. They were regarded as the most important molecules of the living cell, and very little was known about them.

We did know that proteins are made of polypeptide chains. They in turn are made of amino acids, and each amino acid is made of atoms of oxygen, carbon, nitrogen, and hydrogen. Two of them also contain sulphur. It was known that there are approximately twenty different amino acids, but the exact number was not known until much later – not until Watson and Crick guessed it right in the mid 1950s. The important thing is that amino acids can be joined together – it is as if they contained press-studs with male and female parts. One amino acid can join another with the elimination of a molecule of water, leaving male and female press-studs free at either end. In this way many amino acids can join together to form long chains. Along those chains the different amino acids were thought to be arranged in some sort of order, but nobody had any idea what it was.

Emil Fischer, one of the great classic German chemists, first showed proteins to be made up of chains of amino acids, some longer and others shorter. Today, we know that there are enormously long ones. For instance the protein responsible for Huntington's Disease, a terrible neurological disease, has over 3,100 animo acids joined together in one long chain. There are other proteins, like insulin, with only twenty-one joined together in a chain. So proteins were known to be of different size. They were generally regarded as colloids, molecules, of indefinite structure. People's idea was that the chains were just coiled up in some random fashion. But this seemed odd, because, if their structure was random, how could each of the proteins

catalyse (speed up) a specific chemical reaction? It was all very mysterious.

In the late 1920s, James B. Sumner in Canada succeeded in crystallising an enzyme – urease – and a little later, in the 1930s, John H. Northrop, at the Rockefeller Institute for Medical Research in Princeton, crystallised the enzymes that are responsible for digestion. He crystallised pepsin, which is in the stomach, and trypsin and chymotrypsin, the enzymes which digest proteins in the intestine.

Crystallising these enzymes meant that they must be made of molecules which are all structurally identical. However, that idea was looked at with great scepticism until, in 1934, two years before I arrived in Cambridge, Bernal did a crucial experiment with his research assistant, Dorothy Crowford. (She later married Thomas Hodgkin and became famous for discovering the structure of Vitamin B_{12}, for which she got the Nobel Prize. She is one of the great English scientists; but then she was just a very attractive young girl.) They placed a crystal of pepsin, which they got from a biochemist in Sweden, in a glass capillary, in front of an X-ray beam, with a photographic film behind it. What they got on the film was a pattern of spots. So what? That pattern had great significance. The crystals would give a sharp pattern of spots only if the molecules of pepsin had an exact, specific structure and if every molecule of pepsin in the crystal was the exact counterpart of every other. This experiment showed for the first time that an enzyme really has a specific structure such that every atom occupies a specific place, and is in the right place. *And* it meant that, in principle, X-ray analysis could solve that structure – it could find out how the atoms are arranged in a protein molecule.

I didn't know about this result when I started off from Vienna, but I heard about it in Cambridge and obviously I wanted to work on the structure of proteins. But when I got there, Bernal had no protein crystals. At first, I messed around with various projects and had no proper subject for my Ph.D.; but in the following autumn, in 1937, a Cambridge physiologist, Gilbert Adair, gave me some crystals of horse haemoglobin. Adair was another great Cambridge scientist, because he

was the first to determine the molecular weight of a protein correctly. That was haemoglobin. Up to then, nobody even knew how big proteins were. Adair developed a method to determine the molecular weights of very large molecules by using osmotic pressure. From the osmotic pressure of a haemoglobin solution he was able to work out that its molecular weight was 67,000 times the weight of a hydrogen atom – that was a great advance. He also crystallised haemoglobin, and when he heard that I was interested, he gave me some crystals. I put them in front of an X-ray beam and got a beautifully sharp pattern of spots, proudly showed it to all my friends and said: 'Look what a marvellous X-ray diffraction pattern I got from my crystals!' Then they would ask me, 'But, what does it mean?' and I would be embarrassed and change the subject because I had no idea.

What is worse, at that time I had no idea of how to find out what the diffraction pattern meant. The other graduate students in the lab thought it was crazy to try and solve the structure of a molecule with 10,000 atoms, because X-ray crystallography had not yet advanced to the stage of solving the structure of a sugar molecule with only 55 atoms. But I was young and optimistic and I thought the great thing about difficult problems was to have a go at them. Then a terrible thing happened. Austria, my home country, was overrun by Hitler, overrun by the Nazis, and overnight I was turned from a guest to a refugee. (Imagine, if you have come here from abroad, if someone should come up to you and say, 'Your country has been conquered by a foreign power'; and because your grandparents had the wrong religion – my grandparents were Jewish – you could not go back there, because if you did you would be locked up in a concentration camp and might be killed. Imagine what you would feel when you heard this news.) I was shattered, but there were also some disastrous practical consequences for me, because up to then my father had supported my research here. (People nowadays always ask me, 'What sort of a fellowship did you have when you came to Cambridge?' In those days, in the 1930s, there were no fellowships. If I hadn't had a fairly wealthy father I couldn't have come here.) My parents became refugees and my money ran out

and I thought I would have to give up science and find some other job.

Luckily for me, W.L. Bragg had just become Cavendish Professor of Physics. As a young man, he had come here from Australia and had graduated in physics at Cambridge. In the summer of 1912, his father told him that a German – Max von Laue – had discovered that crystals produce X-ray diffraction patterns, but had not been able to discover their meaning. Young Bragg – he was twenty-two then – took an X-ray diffraction picture of a crystal of common salt and worked out how the atoms of sodium and chlorine are arranged. That was the first time that anybody had found out how atoms are arranged in crystals. After that, he worked out how the carbon atoms are arranged in a diamond, and how the zinc and the sulphur atoms are arranged in zincblende, and that was the beginning of X-ray crystallography. Some of his later work he did in collaboration with his father. He and his father got the Nobel Prize for Physics in 1915. Young Bragg got it at the age of twenty-five, a record that has never been broken. He was the father of X-ray analysis, and I was lucky that he came to Cambridge as the head of the physics laboratory. When I showed him my X-ray diffraction pictures of haemoglobin, he was fascinated by the challenge of extending his method of X-ray analysis to the molecules of the living cell. He got a grant from the Rockefeller Foundation in New York which enabled me to carry on my work. This was marvellous for me, because Bragg was the most ingenious and most experienced in that subject, and I thought that, with his help, surely I would be able to solve the problem of protein structure.

In 1949, after various interruptions due to the war, I thought I really had solved the structure of haemoglobin. I thought it was a cylinder, and in the cylinder the protein chains were arranged in a simple zigzag, all in parallel. I thought I had solved the problem – but then my little team was enlarged, and Francis Crick came to join me as my graduate student, to help me with the work on haemoglobin. And the first thing he did was to show that this model, which I was so proud of, was completely wrong. (This can only happen in England. You could not imagine that in Japan or Germany a graduate student would tell his

professor that his model is completely wrong. In England, graduate students are really independent, which is a good thing.)

I therefore had to start all over again; and for some years I was deadlocked. I tried all kinds of methods of solving the problem, but I got nowhere. In the meantime, the problem of protein structure was being attacked by another young man: Fred Sanger. He was a graduate student in the biochemistry department here, and he thought he would try and find out how the twenty or so different amino acids are arranged along a protein chain, and for this purpose he invented new methods of chemistry. (He still lives near Cambridge – a great chemical inventor.) Sanger thought he would attack the smallest protein he could find – insulin – and he invented a method of sticking a dye on to the amino acid at the end of the chain and then splitting it off, and identifying its nature. Having done that, he did the same again – he stuck a dye on to the end of the next amino acid, split it off and identified it. But he couldn't repeat this process right along a long chain, because chemical complications set in. He could only do it with short chains. However, he could split the insulin chains at random into various fragments by boiling the protein in acid: then he fished out the individual fragments, separated them by chromatography and applied his method step by step to each of them. Having done this, how did he know how all the different fragments should be put together? The random splitting gave him a great many different fragments, partially overlapping, so that he was able to tell by the overlaps how they fitted together.

After doing this immensely laborious research – it sounds simple but it took years – he was able to write down the complete sequence of the twenty-one amino acids along the first of insulin's two chains: insulin contains a second chain with thirty amino-acids, and finally there are sulphur atoms that form bridges between the chains. After many years' work, Sanger was able to determine the complete chemical formula of insulin. That was the first time the chemical formula of any protein had been determined. It caused a sensation, because it proved for the first time that protein had a specific arrangement of amino acids along its chain. By determining the sequence of the

different amino acids in the insulin molecules of different animals, Sanger was able to show that the sequence is genetically determined. It was a fantastic breakthrough and earned him the Nobel Prize for Chemistry in 1958. Sanger was a great inspiration to us, because before that, even if we had solved a protein structure by X-ray crystallography, we wouldn't have had the chemical knowledge to build an atomic model, but now we knew that in principle, at any rate, this was possible.

I was still struggling with the problem of interpreting my X-ray diffraction pictures of proteins when another young man arrived. One day in 1951, the door of my lab opened and a man with a crew-cut and bulging eyes popped his head in, without so much as a 'hello', and said, 'Can I come and work here?' I did some quick thinking and remembered that my colleague John Kendrew had had a letter from an American biologist, Salvador Luria, telling him that young Jim Watson wanted to come and join us in Cambridge, and this was he. In his famous best-seller, *The Double Helix*, Watson describes himself as a kind of western cowboy entering our genteel circle in Cambridge. In fact his arrival had an electrifying effect on us, because he told us that more important even than the structure of proteins was that of DNA, because genes are not made of protein, but of deoxyribonucleic acid, or DNA. That discovery had been made at the Rockefeller Institute for Medical Research in New York by Avery, Macleod, and McCarty, and published in 1944. We knew about it, but somehow its tremendous importance hadn't sunk in. The great riddle was, if genes are made up of DNA, how is the genetic information laid down, how is it copied and transmitted from one generation to the next? Watson persuaded Crick to drop his work on haemoglobin and instead help him to determine the structure of DNA.

DNA consists of a chain of sugar molecules, five-membered rings made up of carbon, oxygen, and hydrogen. Successive sugars are linked together by phosphate groups which form bridges between them. Just as the amino acids have different side-chains, so do nucleic acids. The side-chains are bases made up of rings of carbon and nitrogen and they

are of four different kinds. We can think of them as single letters: T, G, C, and A. They form the genetic alphabet – the code in which the genetic language is written.

Watson and Crick attacked the problem by trying to build an atomic model. They knew what the X-ray diffraction pictures of DNA looked like, because Maurice Wilkins and Rosalind Franklin at King's College in London had taken X-ray diffraction photographs of fibres of DNA which gave some indication of the kind of geometry the DNA molecule has. By ingenious reasoning about the properties of long chain molecules and the nature of the X-ray diffraction pattern, Watson and Crick were able to guess the correct atomic model. One Monday morning they called me into their room, which was next to mine, to show it to me (see FIGURE 18.1).

It looked like nothing on earth. They had fixed it to retort-stands reaching from floor to ceiling. They had represented the bases by aluminium plates, and the sugar rings and phosphates by rods of brass, all held together with retort clamps. Two nucleic acid chains were coiled around each other in a double helix, looking like a spiral staircase in which the phosphate–sugar chains form the banisters and the four different bases the steps. The sequence of bases in the two helices had to be complementary, so that A was always joined to T, and G to C. This was the model's most vital feature.

What did it mean? The marvellous thing about it was that it told us what the genetic information consists of and suggested a way in which that information might be replicated every time a cell divides, so that it is transmitted accurately from parent to daughter cells.

The model suggested that when a cell divides, the two chains of the double helix separate and each becomes a template for the growth of the complementary chain along it. In this way, one double helix with a specific complementary sequence of bases is replicated to form two daughter double helices, each with the same complementary sequence of bases as the parent double helix, thus transmitting the genetic information from parent to daughter cells. This simple idea forms the secret of almost all life – not quite *all* life because some viruses carry another

kind of nucleic acid, RNA – but the genetic language which specifies how these viruses are made is the same. All the information which specifies *us*, the information which specifies a human being, is laid down in forty-six chromosomes, which together contain a length of DNA of about one metre. This one metre contains about three billion bases, the equivalent of a library of about 5,000 volumes, all packed together into a single cell. Because the DNA is of atomic dimensions, that information is packed so tightly that it can fit into a single sperm or a single egg.

Watson and Crick's was a fantastic discovery. It was made in March 1953. This was a wonderful year, because in July of that same year, at last, I discovered a way of solving the riddle of the X-ray diffraction patterns of haemoglobin. We attached two atoms of mercury to each haemoglobin molecule. I then compared the X-ray diffraction pattern of this mercury–haemoglobin crystal with that of another crystal that was naked, mercury-free. In principle, this trick solved the problem that had preoccupied me for sixteen years, but there were still a lot of technical complications to be overcome before I solved it in practice.

I have not mentioned John Kendrew yet. He joined me after the war. He had been in operational research in Sri Lanka, and there he had come across Bernal who enthused him about proteins and made him decide to switch from physical chemistry to X-ray crystallography. When the war ended, Kendrew came to Cambridge and joined me to work on the structure of proteins. He thought that haemoglobin was too big a molecule and decided to work on the simpler molecule of myoglobin, which is the haemoglobin of muscle. (Your steak is red not because it contains blood but because it contains myoglobin.) It is a molecule a quarter the size of haemoglobin and it also binds oxygen. After my experiment with mercuri-haemoglobin, Kendrew and his collaborators attached heavy atoms to myoglobin, which turned out to be easier than attaching them to haemoglobin. In 1957, Kendrew built the first rough model of myoglobin. It was a horrible, visceral-looking object, because Kendrew had used a long sausage of plasticine to

represent the protein chain, bent it round as in myoglobin, and fixed it on to wooden sticks from a children's building kit. The sausage mapped out the course of the protein chain. Kendrew's discovery was sensational, because it showed for the first time that X-ray crystallography can really deliver the goods, that it can solve protein structures, which only very few people had believed.

Two years later, Kendrew built the first atomic model of myoglobin. He built a forest of $1/8''$ steel rods; on to the rods he fixed meccano clips in a colour code that indicated density. At points of high density he placed atoms, and the sequence of atoms marked out the amino acids of the protein chain.

That model was the result of an enormously long calculation – a calculation that could not possibly have been done by hand; but by very good luck we had here in Cambridge one of the first two digital electronic computers. (You may think that IBM in the United States were the first to build one, but that was not so. The first working digital electronic computers were the one built here in Cambridge and another in Manchester.) Thanks to that pioneering work of the Cambridge University Mathematics Laboratory, Kendrew and I had the computing capacity to solve the first protein structures. The computer had electronic valves, and its output took the form of punched tape. There was a huge room in the Mathematics Laboratory which was filled with cabinet after cabinet of hundreds of thermionic valves producing an enormous amount of heat; great fans were necessary to extract it. (Nowadays, a computer of the same power is about the size of my briefcase.) This computer had to work all night, or rather the two men who helped Kendrew – Bror Strandberg from Sweden and Dick Dickerson from America – worked all night. In the morning we saw the output of the computer on miles of punched tape. The output was then plotted on paper. It took the form of contour maps specifying the atomic density in the myoglobin molecule in a series of sections, like the microtome sections through a tissue, but on a thousand times smaller scale. These maps allowed Kendrew to build the model. At the same time, I got a rough model of haemoglobin which showed four chains

of high density all coiled in a similar way to the single chain of myoglobin.

I have not yet told why I worked on haemoglobin, and what it does. It is the protein of the red blood cell. It carries oxygen from the lungs to the tissues and helps the return transport of carbon dioxide to the lungs. It is a protein that evolved at the same time as large animals, because without this oxygen-transport molecule, the life of the higher animals would not be possible. So it is a crucial protein. If you are anaemic you are short of haemoglobin, you don't get enough oxygen and feel weak. There are 250,000,000 red cells in one small drop of blood, and inside every red cell there's the same number – about 250,000,000 – of haemoglobin molecules. Haemoglobin contains four atoms of iron embedded in a dye – a pigment – called haem; this is what makes blood red. The four haems in the haemoglobin molecule are arranged in separate pockets on its surface.

Discovering its structure was wonderful. You must imagine the time when proteins were black boxes. Nobody knew what they looked like. There I was, having worked on this vital problem for twenty-two years, trying to find out what this molecule looked like, and eventually to discover how it worked. When the result emerged from the computer one night and we suddenly saw it, it was like reaching the top of a difficult mountain after a hard climb and falling in love at the same time. It was an incredible feeling to see this molecule for the first time and to realise that my work had not been in vain: because at many stages during those long years I feared that I was wasting my life on a problem that would never be solved. It was marvellous that it came out, but that was only the first stage, when we saw how the four chains were coiled. The real thing was to find how the atoms were arranged and how the molecule works. Haemoglobin is not just an oxygen tank: it is a molecular lung. It changes its structure every time it takes up and releases oxygen. That change of structure is vital for its oxygen–carrying function. You can hear your heart going 'thump, thump, thump', but in your blood the haemoglobin molecules go 'click, click, click' all the time – but you can't hear that! Another exciting thing was

that the way in which the chain was coiled in haemoglobin turned out to be the same as in myoglogin. We soon realised that the fold is universal in nature, and is the way all haemoglobins are built.

Further reading

Max Perutz. 1990. *Mechanisms of Cooperativity and Allosteric Regulation in Proteins*. Cambridge: Cambridge University Press.

Max Perutz. 1992. *Protein Structure. New Approaches to Disease and Therapy*. New York: W. H. Freeman.

Max Perutz. 1997. *Science is Not a Quiet Life*. London: Imperial College Press, World Scientific.

Francis Crick. 1988. *What Mad Pursuits*. New York: Basic Books.

John C. Kendrew. 1966. *The Thread of Life*. London: C. Bell & Sons.

Max Perutz was born in Vienna in 1914. He studied chemistry at the University of Vienna until, in 1936, he moved to the Crystallography Department at the Cavendish Laboratory in Cambridge where he began his work on the structure of haemoglobin, then a seemingly insoluble problem. In 1953 he found a way of solving it which opened the field of protein crystallography and led to the first model of the molecule in 1959. At the same time, his colleague John Kendrew solved the structure of the related, smaller molecule of myoglobin. For these discoveries they received the Nobel Prize for Chemistry in 1962. In 1947, Perutz founded the Medical Research Council Unit of Molecular Biology which developed into the Laboratory of the same name in 1962. He became its first Chairman until his retirement in 1979. He still works there now.

22 The discovery of pulsars – prelude and aftermath
ANTONY HEWISH

The early history of the Cavendish Laboratory is best known for ground-breaking work in the study of atomic structure, but the birth of long-distance radio communication also stimulated research on the properties of the upper atmosphere which enabled such propagation to occur. Begun by E.V. Appleton, who was later awarded the Nobel Prize in 1947 for his discovery of the ionosphere, radiophysics at the Cavendish was continued until 1939 by J.A. Ratcliffe and it was he who initiated radioastronomy in 1945. What triggered Ratcliffe's interest was an occasion in February 1942 when radar stations along the south coast were blinded by radio interference, initially thought to be jamming by enemy action, but later found to be radiation emitted by the sun when a large sunspot was present on the disk. Anxious to regenerate radiophysics at the end of the war, Ratcliffe attracted M. Ryle, a wartime colleague, back to the Cavendish and suggested that investigation of this new solar phenomenon might be an interesting project.

By 1946 Ryle had set up a primitive radio telescope and discovered that the sun was a continuous emitter of radio waves, in addition to the more intense outbursts associated with sunspots. More importantly, however, he demonstrated the existence of other celestial radio emitters, then called radio stars, and radioastronomy in Cambridge had begun. I joined Ryle's group in 1948 and this essay outlines the course of my personal research, leading to the discovery of pulsars in 1967, which has been ranked as one of the major astronomical breakthroughs of the past fifty years. The history of science is punctuated by unexpected discoveries and it was my good fortune that the rather unusual requirements of my programme dictated the design of a radio telescope that was ideally suited for detecting the totally unpredicted radiation from this exciting new class of stars.

My first encounter with Ryle was in 1944 at the Telecommunications Research Establishment, Malvern. Ryle was designing airborne devices to jam enemy radar systems and my task was to instruct RAF personnel in the use of the equipment so I joined his team for a few months prior to visiting the RAF Bomber Command station where it was first installed. I had not planned to do research on completing my interrupted degree course in 1948, but to my tutor's surprise I got a first in the Natural Sciences Tripos and was offered a research studentship at the Cavendish. Ratcliffe then told me about Ryle's work – at that time I was unaware that he had returned to Cambridge. Having seen a few other research groups, and with my wartime background, it became clear that joining Ryle was the obvious choice.

The immediate goal was to survey the sky at higher sensitivity to locate a larger number of radio stars. Were they similar objects to the sun but more distant, or were they something quite new? Certainly their positions did not coincide with any of the brightest visible stars. A larger radio telescope was essential so I helped with the construction and design of a dipole array. Few laboratory assistants were available so everyone joined in the work of hammering iron posts into the ground and I became adept with a hacksaw after cutting stacks of brass-tubing into lengths for the dipoles. As a more cerebral task Ryle asked me to investigate variations of intensity shown by some of the radio stars which suggested some similarity with solar radio emission. Analysing past recordings, I noticed that the variations were strongest when each radio source was under observation near midnight. Following up this midnight effect I found a close connection between the radio variations and the occurrence of a type of disturbance in the upper ionosphere. This confirmed that the variations were a local effect, like the twinkling of visible stars caused by atmospheric turbulence.

My discovery of a close link between radioastronomy and the ionosphere delighted Ratcliffe who was involved with purely ionospheric research. It also intrigued me because it opened up the possibility of studying regions of the ionosphere inaccessible to the ground-based methods then in use. The ionosphere is formed when

ultra-violet radiation from the sun strips electrons from atoms, and radio waves transmitted from the ground are reflected near the middle of the ionised layer where the electron density is highest. The shorter wavelengths used in radioastronomy pass right through and give information at greater heights above the level of reflection. Exploiting this new method became the main topic of my Ph.D. research and essentially defined the course of most of my subsequent work.

Initially I had to develop the necessary theory and here I was much assisted by graduate lectures given by Ratcliffe. He was a superb teacher with a special interest in the application of Fourier analysis to physical problems. Several of my contemporaries look back to those lectures as a landmark in their scientific training. Aided by that grounding I calculated how clouds of electrons in the ionosphere refracted the incoming radio waves and then used this to determine the characteristics of the clouds from our observations. The situation is similar to that when sunlight falls on the wavy surface of a swimming pool and causes a mobile tracery-pattern of intensity on the bottom. Tracking these patterns enabled me to measure wind speeds at heights of several hundred kilometres.

My work formed a nice bridge between the groups of Ryle and Ratcliffe, but when both the theoretical and observational aspects were being developed further by new members of Ratcliffe's team it was time for me to move back to astronomy. At this stage the sun beckoned once again. Traditionally the sun's atmosphere, the solar corona, could only be studied during a total eclipse when blinding light from the solar surface is blotted out. It had, however, been suggested that radiation from a radio source observed at a small angle from the sun might be refracted by the corona, the hot ionised gas acting like a giant, spherical diverging lens. Successful observations were made with unexpected results. Instead of the sudden cut-off of intensity due to lens action there was a slow decrease over several days. This could be explained if the ionised corona contained clouds of varying density giving irregular refraction. My theory for the ionospheric clouds was directly applicable and I successfully measured the size of the clouds.

FIGURE 22.1 Part of the 4.5 acre array at the Mullard Radioastronomy Observatory which was used in the discovery of pulsars.
Source: Cavendish Laboratory Archives.

Over the next few years, with my own students, we studied the solar corona progressively further and further from the sun. In 1956 radioastronomy moved from its original site on the old University Rifle Range beside the Coton footpath to Lord's Bridge, beyond Barton on the A603. Then in 1964 came a development which added a new dimension to my work and formed an essential prelude to the discovery of pulsars three years later. Ryle's pioneering sky surveys had soon established that the term radio star was a misnomer. In reality most of the celestial radio emitters were galaxies of a new kind located far beyond the visible galaxies. We now know that the radio-emitting galaxies contain compact, massive central nuclei, probably giant black holes, which generate prodigious energy as galactic material is sucked inwards by gravitational forces. It was found in 1962 that some radio galaxies were quasars (quasi-stellar galaxies), characterised by point-like regions of intense radio emission about one hundred times smaller than typical radio galaxies. It turned out that quasars 'twinkled' because of the clouds in the solar atmosphere. I immediately saw two exciting possibilities for further research. Just as I had measured ionospheric winds by observing the speed of radio intensity patterns sweeping over the ground it should similarly be possible to study solar winds in space. Secondly, the degree to which a radio galaxy scintillated gave an indication of the presence of an active nucleus and could be used to identify further quasars. This was important in the mid-1960s, long before radio telescopes could reveal the structure of radio galaxies.

With my student P.A. Dennison we began with the solar wind. The problems here were the very high speed, about one million miles per hour, and the large scale of the intensity pattern characterised by patches about one hundred miles across. We needed simultaneous observations at sites at least thirty miles apart. I designed simple aerial systems, locating one on Viscount Elveden's estate near Thetford, another near Clacton on the coast, and the third at Lord's Bridge. I hired telephone lines to return the data. Only a few quasars could be observed but it all worked well and our measurements agreed nicely with in-situ observations from the earliest spacecraft outside the

earth's atmosphere. One advantage, as compared to direct measurements in space, was our ability to measure winds blowing from the sun's polar region. This region was inaccessible to spacecraft due to the difficulty of achieving a suitable orbit passing over the solar pole. We found that the polar wind was twice as fast as that from lower solar latitudes and it was not until 1994 that this was finally confirmed by the spacecraft Ulysses.

Our pioneering work on the solar wind was later continued by groups in the USA and Japan, but by then we had discovered pulsars. Exploiting scintillation to identify quasars demanded a radio telescope of unusual design. To detect about 500 radio galaxies needed high sensitivity which required a surface area of several acres, and repeated observations were essential to find how the scintillation varied with the angle of the line of sight from the sun. Pointing a conventional radio telescope towards each radio galaxy in turn would have taken far too long. So I designed something akin to a fly's eye which can look in many directions at once. It was an array of 2048 dipole aerials arranged in lines over an area of 4.5 acres. Jocelyn Bell from Glasgow University joined me as a research student in 1965 and worked hard in the construction team with special responsibility for the cable network connecting the dipoles. The array was completed in July 1967 and the sky survey commenced. Observing in four directions simultaneously the full sky survey took four days and this was repeated each week to measure scintillation on each radio galaxy at varying angles from the sun. Jocelyn needed the results for her Ph.D. so she ran the survey and became adept at analysing the charts.

Man-made radio interference is one of the bugbears of radio astronomy. Anticipating that interference would be a serious problem I requested Jocelyn to plot the position of every fluctuating signal on a sky map. As the data accumulated genuine scintillating radio galaxies would be repeatedly plotted in the same position while interference would give points scattered randomly. A clear distinction between real quasars and interference was usually possible but one day in late August (neither of us can recall the exact date) Jocelyn pointed out

some odd recordings. One source had appeared in the same position several times, giving a somewhat scruffy trace, but occasionally it was absent. Another puzzling feature was the high level of scintillation at such a large angle from the sun where scintillation from other quasars was weak. I first thought that we had discovered a radio-flaring star, similar to the sun but at a much greater distance. I checked with my colleague David Dewhirst at the Cambridge Observatory but he knew of no flare star in that position. To gain more information I asked Jocelyn to use a more sensitive recorder running at high chart speed so that we could see the fluctuations more clearly. Around lunch time on 28 November, just after I had finished lecturing, she phoned to say that the fast record showed pulses of varying height at intervals of just over one second. I thought that such an unnatural-looking signal could only be man-made radio interference. Next day I was not lecturing and saw the pulses for myself as they were again recorded.

Early December in the 1960s was a hectic time for dons involved with College scholarship examinations and all the next week I was confined to my room in Churchill marking scripts. At lunch one day I mentioned the pulses to the distinguished geophysicist Edward Bullard and he, quite seriously, suggested the possibility of aliens. Jocelyn and I, tongue in cheek, had already nicknamed the signal LGM (Little Green Men)! With examination work completed, I could spend more time at the observatory and on 10 December I began to time the pulses accurately, using the seconds time-pips broadcast continuously from Rugby (The national MSF time service). To my amazement the periodicity was maintained to at least one millionth of a second. This ruled out most possibilities of man-made interference, except perhaps, for equipment operated by other astronomers. I checked locally, and also with the Royal Greenwich Observatory, and drew a blank. Perhaps some distant spacecraft was responsible? One important factor here was some changes in position, up to one third of a degree, which Jocelyn had reported in the week to week survey records. A genuine astronomical source could not behave like this. But Jocelyn had been using the time of maximum intensity to obtain her position. I

tried another method, exploiting the fact that I had designed the array in two halves, connected as a simple interferometer. The phase reversals, which are a characteristic feature of this arrangement, denoted a more precise position and when I reanalysed all the data using this method I found no change whatever in the source location. Rapid variations of intensity had produced an apparent position-shift. The signals therefore had to be coming from far outside the solar system. With John Pilkington, another member of my team, we next found that each pulse was caused by a signal sweeping from higher to lower radio frequency through the operational waveband of the radio telescope. This is to be expected for a wide-band pulse that has travelled for about one hundred years through the interstellar gas, and it indicated an origin amongst the nearer stars. In addition we knew, from the short duration of the pulses, that the emitter could not be larger than a small planet. The possibility that aliens could have generated such an unnatural-looking radio signal was no longer a joke. I discussed with Ryle how we should handle the discovery if the alien theory could not be dismissed and we agreed that presentation of our observations to the Royal Society for comments and advice from top scientists in the country would be the best procedure. No hint of the discovery was given to anyone outside the team as any leakage would have triggered an unwelcome invasion from the media. I was determined to maintain a cool scientific atmosphere until we had completed all possible measurements, and crucial evidence was still to come from my day-to-day timing of the pulses. Within a week I had detected small changes in the pulse rate due to the curved orbit of the earth around the sun, and a similar effect would occur if the source itself was located on an alien planet circling some distant star. I plotted the daily measurements with some apprehension until, by early January 1968, the absence of any additional Doppler shift showed that the putative alien planet had to be so far from a star that the conditions for life were most unlikely. In parallel with my work at the observatory, Jocelyn had been reanalysing all past survey records in a search for additional pulsing sources, and further observations with the fast recorder confirmed the

existence of three more. This was additional evidence against the alien theory.

By mid January the possibility of aliens had receded and, with some relief, I began to write an account of our remarkable discovery whilst also searching the literature on midget stars for some plausible explanation of pulsed radiation. Cepheid variables are well-known stars which periodically expand and contract, accompanied by corresponding variations of brightness on time scales of several days. The periodicity, controlled by gravity, depends on the density of the stellar material and to explain our pulses would need a density more than ten times that of white dwarf stars, the most compact stars then known. One remaining possibility was the hypothetical neutron stars, over a million times more dense than white dwarfs and crushed by their enormous gravity into spheres only ten miles in diameter. Their existence had been surmised as a theoretical possibility after Chadwick's discovery of the neutron in the 1930s, but was still highly speculative. Remembering the powerful bursts of radio emission generated by shock waves from eruptions in the solar atmosphere, I ended our paper with the suggestion that the pulses might be caused by analogous disturbances driven by periodic vibrations of neutron stars, or possibly white dwarf stars.

The journal *Nature* published our discovery within two weeks but I wanted local astrophysicists to know about it first, so a lecture was arranged at the famous Maxwell Lecture Theatre in the Cavendish Laboratory, which was packed. As the news spread it met with scepticism in some quarters; Philip Morrison, science correspondent of the *New York Times*, later told me that he could not believe that our observations were genuine – we must have been fooled by some artefact. Other radio telescopes, including the Lovell Telescope at Jodrell Bank, soon confirmed the pulses however, and excitement spread around the world triggering a flood of theoretical speculation. The signals were not hard to find, knowing what to look for, and it was our good fortune that the observational requirements of the scintillation survey happened to be so well suited to the initial detection.

Ultimately, the most satisfactory theory was due to Thomas Gold, a Cambridge astrophysicist who had moved to Cornell in the USA. His idea was that a spinning neutron star generated a beam of radio waves, thus flashing regularly at the rotation period. This 'lighthouse' model was dramatically confirmed in 1969 by the discovery of a pulsar near the centre of the Crab Nebula, the famous remnant of a supernova seen to explode in the year 1054. It had been postulated in the 1930s that neutron stars might be formed by gravitational collapse when massive stars ran out of nuclear fuel. The protons and electrons would then be crushed together to form neutrons, while neutrinos released in the same process would initiate a massive explosion. The pulsar within the Crab Nebula was flashing very rapidly at thirty times per second and was found to be slowing down at precisely the rate expected for a neutron star which had been created during the explosion. Moreover, only a neutron star could survive rotating at such high speed without ripping apart under centrifugal force.

It might appear that neutron stars would be objects of little interest, being composed of the compressed nuclear waste of burnt-out stars. In fact, because of the enormous shrinkage, by a factor of one hundred thousand in linear size, there is a huge gain of rotational energy which provides a source of power and endows neutron stars with a fresh lease of life. Far from a dull old retirement, they embark upon a sprightly new career. Another vital factor in this rebirth is the intensification of their intrinsic magnetisation by up to one hundred million times. A spinning magnet is, of course, a dynamo, so a neutron star generates an ultra-high voltage across its surface. Charged particles are therefore accelerated to extremely high energies by the electrical forces and ejected from the star. Direct evidence of this can be seen in the glow of white light and emission of X-rays from the Crab Nebula. It had been known for many years that this radiation came from charged particles weaving at high speed through the tangled magnetic field of the supernova remnant, but the origin of the particles was puzzling since they needed to be continually replenished. Identification of the central neutron star solved that problem.

Even after thirty years of research, including sky surveys which have now catalogued over one thousand pulsars, we still do not really know how the 'lighthouse' beams of radio waves are produced. Evidence from the polarisation of the radiation, and the angular spread of the beams, favours the theory that high-speed outflows of accelerated particles, guided along the magnetic field from the magnetic poles, develop wave motion like that due to wind-shear in our atmosphere as occasionally revealed by narrow bands of cloud at regular spacing. The stellar wind is composed of charged particles and the waves cause oscillating electric currents to flow so that antenna-like radiation is generated. The high speed directs the radio beam along the magnetic axis and if this is oblique to the axis of rotation a lighthouse beam is obtained.

Apart from their astronomical interest pulsars provide fascinating arenas for the study of matter under extreme conditions far removed from anything reproducible in terrestrial laboratories. It is hard to imagine matter so compressed that one teaspoonful would contain enough material to build one hundred of the largest ocean liners, yet this is what the inside of a neutron star is like. From the known properties of neutrons it was predicted, remarkably, that neutron stars should contain a most unusual type of liquid – a quantum liquid. The bulk motion is constrained by the laws of quantum physics, and the particle-waves which normally relate to individual particles, as in an atom, now bind the neutrons en-masse. This liquid cannot circulate, as in a stirred cup of coffee, such motion being forbidden by quantum laws. When neutron stars form they are, of course, spinning rapidly and the liquid rotation is compacted into a forest of tiny, thread-like vortices aligned parallel to the rotation axis. These quantised vortices terminate on the outer shell of the neutron star which is a rigid layer of compressed, iron-like material. Careful timing of some pulsars over a number of years has revealed effects which confirm this strange interior structure. The gradual slowing down, as the star loses energy into outflows of particles and radiation, is occasionally interrupted by sudden, tiny increases of pulse rate indicating a faster spin of the outer shell. This can happen if some of the mini-vortices in the liquid

dissipate by transferring their spinning motion to the shell. Calculations have shown that the sudden changes of pulse rate can be explained by this behaviour.

Most pulsars do not exhibit changes of pulse rate, apart from their systematic slowing down, and even this is very small. The time-span required for the pulse rate to decrease substantially is typically a few million years. This means that they are, in effect, extremely good clocks and they provide time standards comparable to the best atomic clocks. This has been exploited in various ways. Just as I used timing data to check whether the signals were coming from an alien planet in orbit about some distant star, one pulsar has a pulse rate which varies periodically, indicating that the neutron star has three planets in orbit around it. Their environment could not support life, but this was the first definitive evidence that other stars can have planetary systems. The binary pulsar discovered by J.H. Taylor and R. Hulse in 1975 has also been very important. Here a pulsar is in a close orbit about another neutron star with an orbital period just over seven hours and the companion to the pulsar is, as yet, undetectable. What makes the binary pulsar exceptionally interesting is the steady decrease in separation of the neutron stars. Continued timing since 1975 has shown that this distance is shrinking by about three metres per year. Three hundred million years from now the neutron stars will collide and coalesce. The steady loss of energy associated with this orbital collapse is in exact agreement with Einstein's theory of general relativity which predicts that gravitational waves must be launched from massive orbiting systems. The four-dimensional spacetime which defines our physical world is warped by gravitational fields and the essence of Einstein's theory is that gravitational forces are explained by these distortions. Dynamical arrangements like the binary neutron stars generate ripples in spacetime travelling outwards at the speed of light and also carrying energy. Taylor and Hulse were awarded the Nobel Prize in 1994 for confirming, indirectly, the existence of gravitational waves through their discovery and subsequent research on the binary pulsar.

Searches for the most rapid pulsars have found some rotating at more than one hundred revolutions per second. These are believed to be neutron stars which originally formed when one member of a binary pair evolved to become a supernova. If both stars remain in orbit after the supernova explosion the neutron star may be close enough to its normal stellar companion to accrete matter via gravitational attraction. This will spin-up the neutron star like a whip-top giving the high spin rates observed. Just over 600 revolutions per second is the highest rate yet found. If pulsars much faster than this turn up they could not be neutron stars because even these disrupt at such high speeds. More compact, exotic stars composed of quarks or other particles have been conceived, but particle physics is not yet able to describe what these would be like. Even after thirty years the pulsar story is far from over.

Further reading

The Astronomy Encyclopaedia. London: Mitchell Beazley International.
Jayant V. Narlikar. 1999. *Seven Wonders of the Cosmos*. Cambridge: Cambridge University Press.
Andrew G. Lyne and Francis Graham-Smith. 1998. *Pulsar Astronomy*. Cambridge: Cambridge University Press.

Antony Hewish (1924–) Radio astronomer, born in Fowey, Cornwall, SW England, UK. He studied at Gonville and Caius College, Cambridge, from 1942–3 and spent three years war service at the Royal Aircraft Establishment, Farnborough. Returning to Cambridge in 1946 he began research at the Cavendish Laboratory in 1948 and used the scintillation (twinkling) of radio galaxies to study clouds and winds in the ionosphere, later extending the same technique to measure the solar wind in space. In 1965 he designed a novel radio telescope for utilising scintillation to study quasars and in 1967 this led him and his student Jocelyn Bell-Burnell to discover the first pulsars, rapidly spinning neutron stars, which opened a new chapter in astrophysics. For this discovery he was awarded the Nobel Prize for Physics in 1974, which he shared with his former teacher, Sir Martin Ryle. Appointed Professor of Radioastronomy in 1971 Hewish became Director of the Mullard Radio Astronomy Observatory, Cambridge from 1982–87. As Fellow, and Director of Studies in Physics at Churchill College from 1961–71, Hewish was much involved with teaching and as Professor of the Royal Institution, London, from 1977–79 he was keen to foster the public understanding of science.

23 Stephen W. Hawking
SIMON MITTON

British theoretical physicist and cosmologist, Stephen Hawking is world-renowned for his discoveries on the nature of space and time, and as the author of A Brief History of Time, a science book with sales that broke many publishing records. Hawking achievement's are all the greater in view of his crippling illness which has left him without speech and only limited movement in his hands.

Hawking was born in Oxford, England, on 8 January 1942, precisely 300 years after the death of Galileo. He says of this remarkable coincidence, 'I estimate that about two hundred thousand other babies were also born that day, and I don't know whether any of them became interested in cosmology'. He went to local schools, was slow at learning to read, and never stood out as a scholar. His father, a doctor, wanted Stephen to study medicine at university, and as a result the young Hawking studied very little mathematics in high school. However, he gained a place to study physics at Oxford University, where he was admitted to University College in 1959.

Stephen Hawking describes the Oxford of those days as being, 'Very anti work – you were supposed either to be brilliant without effort or accept your limitations and get a poor classification in the final examinations'. Hawking claims to have done no more than 1000 hours of studying in his entire three years at Oxford, and he attributes this lack of effort to an attitude of complete boredom and a feeling that nothing was worth the effort. He graduated with the highest honours in his final exams.

The deep-rooted ennui changed dramatically soon after he went to Cambridge University (1962) to work for a doctorate in general relativity. He had started to become clumsy in his movements. The diagnosis of this condition was amyotrophic lateral schlerosis (motor

neurone disease), an incurable disease, which progressing affects the brain functions controlling movement. At first the disease progressed rapidly, and Hawking believed he would not live long enough to finish the doctoral programme. On the point of giving up studies, he fell in love with Jane Wilde, and has repeatedly said that, 'she gave me something to live for'. They soon married (1965), and Hawking went back to research with renewed enthusiasm as he needed to get a job.

From the beginning Hawking's research has been on singularities in the universe. A singularity is a mathematical concept. It can be visualised as a region of spacetime which has become so highly curved that normal physical quantities become infinite and the ordinary laws of physics cease to apply. Spacetime is the four-dimensional grid used in theoretical physics to locate events and describe the spatial and temporal relationships between them. A famous example of a singularity in spacetime is the Big Bang event that gave rise to the expanding universe we observe today, and Hawking's early research was on this.

Observations of distant galaxies show that the universe is expanding, which implies they were closer together in the past. One can then ask if there was a time when all the galaxies were on top of each other, and the density of the universe infinite. To answer this required new mathematical techniques which Stephen Hawking developed with Roger Penrose from 1965–70. Together they showed that the universe must have been infinitely dense in the past: this is the Big Bang singularity, the beginning of our universe.

In the 1970s Hawking turned to another type of singularity: collapsed stars. The general theory of relativity predicts that when a star has exhausted its nuclear fuel, it will collapse under the crush of its own gravity. Although stars like the Sun can form stable white dwarfs, Hawking showed that more massive stars would continue to collapse until they reached infinite density: this singularity would be the end of time, at least for the star and anything on it. The gravitational field would be so strong that nothing could escape from the singularity. Such a region is called a black hole and its boundary is called the event horizon.

Continuing his research on black holes, Hawking found that the surface area of the event horizon could not decrease with time. In a flash of insight one night Hawking realised that if two black holes collided the surface area of the new black hole would be larger than the sum of the surface areas of the original black holes, 'I was so excited I did not get much sleep that night'.

BLACK HOLES ARE NOT BLACK!

From 1970–4 Hawking worked on black holes with several collaborators. With Brandon Carter and Werner Israel he proved the 'No Hair Theorem'. This states that the only properties of matter that are conserved when it falls into a black hole are the mass, the angular momentum, and the electric charge: all other properties such as shape, chemical composition, and even whether it is matter or antimatter, are destroyed.

In 1974 Hawking came up with the astonishing conjecture that black holes are not really black! By applying the laws of quantum theory to a black hole, he showed that in the quantum world – on the very smallest scale of physical reality – particles and radiation leak out of the black hole. This is a consequence of the Heisenberg uncertainty principle and the creation of 'virtual particles' which are fleetingly created near the event horizon. The radiation of heat from black holes and their eventual evaporation is known as the 'Hawking effect', and eventually the black hole explodes in a final burst of radiation.

THE THEORY OF EVERYTHING

In the last twenty years, Stephen Hawking has worked almost exclusively on a theory of everything. There are four fundamental forces in physics. These are gravity, which is described by general relativity, the electromagnetic force, and the strong and weak nuclear forces. Physicists have already succeeded in producing a single theory that embraces electromagnetism, the strong force and the weak force. In the very early universe, up to the time when our universe was one millionth of a millionth of a second old (10^{-12} seconds) these three forces

were unified into a single quantum force. The goal is to find a consistent physical theory in which general relativity and the quantum laws can be united. This would yield a full quantum theory of gravity. Within this single physical theory would be an explanation for gravity, electromagnetism, and the nuclear forces. Furthermore, constants of physics, such as the velocity of light or the strength of the gravitational interaction, would be calculated from first principles, whereas at present they are known only by observation and measurement.

The search for quantum gravity has taken Hawking almost to the beginning of time. It is already clear that the theory of quantum gravity provides the rule book governing the operation of the universe from time zero to just 10^{-43} seconds after the Big Bang. In this interval all the fundamental properties of spacetime, the nature of the universe, and the nature of physical laws are defined.

One result of this investigation has been the proposal, jointly with James Hartle of University of California, Santa Barbara, is that space and time are both finite in extent, but they do not have any boundary or edge. In three dimensions we are familiar with the surface of the Earth, which is finite, but this surface has no edges. So in the compact four-dimensions of spacetime in the early universe, the distinction between space and time is blurred, and there are no edges. According to Hawking this means that the universe can be described by a mathematical model which is completely determined by the laws of science alone. We do not know, yet, the precise form of these laws which govern the behaviour of the universe under extreme conditions. Hawking and his school aim to find these laws by the end of the millennium.

If a single theory is produced would it enable us to predict everything, and thus bring about the 'end of theoretical physics'? No, according to Hawking, 'Our powers of prediction would still be severely limited, first by the uncertainty principle (which states that certain quantities cannot be predicted exactly) and secondly by the complexity of the equations which makes them impossible to solve except for the simplest cases. We are still a long way from omniscience'.

COPING WITH DISABILITY

Stephen Hawking has a profound effect on all those who work with him. It was a great shock to him to find that he had motor neurone disease. Until 1974 he was still able to feed himself, and his wife ran the family home and brought up the first two children without any assistance. From then until 1980 one or more of his research students lived in the family home to help, but after 1980 it became necessary to have professional nursing help. Stephen caught pneumonia in 1985, and had to have a tracheostomy operation as a life-saving measure. This left him completely speechless. He now has twenty-four hour professional care, paid for by grants from several foundations, and with this help he can still travel to scientific conferences.

Hawking now communicates with the aid of a computer system. A computer is attached to his wheelchair. By using the small amount of movement still present in his hands, Hawking manipulates a handheld switch to move the cursor across the computer screen. Pull down menus enable him to select words and build up a sentence word-by-word on screen. His present speed is fifteen words per minute. The text then goes through a speech synthesiser, which causes him to speak, apparently, with a strong American accent!

The same computer word-processor is used for writing books and even the complex mathematics in his research papers. There is a link to a portable telephone, used by the speech processor to send his outgoing calls. In this way Hawking has made important contributions to improving communications for disabled people in general. He is also a tireless campaigner for better facilities for the disabled, and has openly criticised Cambridge University for inadequate access for persons in wheelchairs.

It is strongly symbolic of his mission to lead as 'normal' a life as possible that he has in recent years campaigned for a political party by appearing in its advertisements on national TV, appeared as himself in a science fiction movie, where he plays chess against Einstein, and has a pin-up of Marilyn Monroe in his Cambridge office.

Further reading

S.W. Hawking and G.F.R. Ellis. 1973. *The Large Scale Structure of Space-Time.* Cambridge: Cambridge University Press.

Stephen Hawking. 1994. *Black Holes and Baby Universes.* New York: Bantam.

Stephen Hawking. Re-issued 1995. *A Brief History of Time.* New York: Bantam.

Laurence Krauss and Stephen Hawking. 2000. *The Physics of Star Trek.* London: Flamingo.

R.S. Penrose and S.W. Hawking. 2000. *The Nature of Space and Time.* Princeton, NJ: Princeton University Press.

K.S. Thorne and S.W. Hawking. 1998. *Black Holes and Time Warps.* London: Papermac.

Michael White and John Gribbin. 1998. *Stephen Hawking, A Life in Science.* London: Penguin.

Jane Hawking. 2000. *Music to Move the Stars.* London: Pan.

Stephen (William) Hawking (1942–) British theoretical physicist: advanced understanding of space time and space time singularities.

Hawking graduated from Oxford in physics and, after a doctorate at Cambridge on relativity theory, remained there to become a Fellow of the Royal Society (1974) and Lucian Professor of Mathematics (1979). He developed a highly disabling and progressive neuromotor disease while a student, limiting movement and speech. His life and work is an extraordinary conquest of severe physical disability.

Simon Mitton studied physics at the University of Oxford, following which he did a Ph.D. in radio astronomy at the Cavendish Laboratory, University of Cambridge. He is Senior Fellow of St Edmund's College, Cambridge.